SpringerBriefs in Applied Sciences and Technology

Computational Mechanics

Series editors

More information about this series at http://www.springer.com/series/8886

Sergey Alexandrov

Elastic/Plastic Discs Under Plane Stress Conditions

Springer

Sergey Alexandrov
Institute for Problems in Mechanics
Russian Academy of Sciences
Moscow
Russia

ISSN 2191-530X ISSN 2191-5318 (electronic)
SpringerBriefs in Applied Sciences and Technology
ISSN 2191-5342 ISSN 2191-5350 (electronic)
SpringerBriefs in Computational Mechanics
ISBN 978-3-319-14579-2 ISBN 978-3-319-14580-8 (eBook)
DOI 10.1007/978-3-319-14580-8

Library of Congress Control Number: 2015936293

Springer Cham Heidelberg New York Dordrecht London

Printed on acid-free paper

Springer International Publishing AG Switzerland is part of Springer Science+Business Media (www.springer.com)

Preface

This monograph concerns with analysis of thin annular elastic/plastic discs of constant thickness subject to mechanical and thermal loading under plane stress conditions. The flow theory of plasticity is used. The presentation of the introductory material and the theoretical developments appear in a text of three chapters. The topics chosen are primarily of interest to engineers as postgraduates and practitioners but they should also serve to capture a readership from among applied mathematicians. The monograph provides both a description of a general approach to finding the distribution of stresses and strains in thin discs and a collection of analytic and semi-analytic solutions. Many solutions are represented by formulae. Such solutions are immediately ready for practical use. Other solutions are illustrated by diagrams. These diagrams demonstrate most important tendencies in solutions behaviour. It is however evident that they cannot be used for practical calculation of stress and strain distributions. Therefore, most of such solutions are described in great detail. In most cases, numerical techniques are only necessary to evaluate integrals and solve transcendental equations. For reasons of space, the main focus is on mechanical and thermal loading, though the general approach can be extended to thermo-mechanical loading without any difficulty.

Among the topics that are either new or presented in greater detail than would be found in similar texts are the following:

1. A general approach to calculate the distribution of stresses and strains in thin annular discs for the von Mises yield criterion and its associated flow rule.
2. A general approach to calculate the distribution of stresses and strains in thin annular discs for Hill's quadratic anisotropic yield criterion and its associated flow rule.
3. A general approach to calculate the distribution of stresses and strains in thin annular discs for Drucker–Prager yield criterion and two flow rules.
4. Analytic and semi-analytic solutions for thin annular discs under specific loading conditions of practical interest.

The first chapter concerns the general approach to calculate the distribution of stresses and strains in thin elastic/plastic discs. In particular, general solutions for stresses and strains are given for three widely used yield criteria. These are the von Mises yield criterion, Hill's quadratic anisotropic yield criterion and Drucker–Prager yield criterion for pressure-dependent materials. In all cases, the associated flow rule is used to connect stresses and plastic strain rates. In addition, plastically incompressible material is considered in case of the Drucker–Prager yield criterion. These general solutions are used in subsequent chapters.

Chapter 2 deals with thin annular discs subject to internal or external pressure. Analytic or semi-analytic solutions are proposed for each of the aforementioned yield criteria.

In Chap. 3, it is assumed that disc inserted into a container is subject to thermal loading. As in the case of mechanical loading, analytic or semi-analytic solutions are proposed for each of the aforementioned yield criteria.

Moscow
March 2015

Sergey Alexandrov

Acknowledgments

The research described in the present monograph has been supported by the Grants RFRB-14-01-93000 and NSH-1275.2014.1.

Contents

Symbols

The intention within the various theoretical developments given in this monograph has been to define each new symbol where it first appears in the text. A number of symbols are introduced in the abstract to individual chapters. These symbols reappear consistently throughout the chapter. In this regard each chapter should be treated as self-contained in its symbol content. There are, however, certain symbols that reappear consistently throughout the text. These symbols are given in the following list.

k	Non-dimensional parameter introduced in Eq. (1.28)
p	Time-like parameter
E	Young's modulus
γ	Thermal coefficient of linear expansion
ε_r, ε_θ, ε_z	Total strains in cylindrical coordinates
ε_r^e, ε_θ^e, ε_z^e	Elastic portions of total strains
ε_r^T, ε_θ^T, ε_z^T	Thermal portions of total strains
ε_r^p, ε_θ^p, ε_z^p	Plastic portions of total strains
ν	Poisson's ratio
ξ_r, ξ_θ, ξ_z	Derivatives of ε_r, ε_θ and ε_z with respect to p
ξ_r^e, ξ_θ^e, ξ_z^e	Derivatives of ε_r^e, ε_θ^e and ε_z^e with respect to p
ξ_r^T, ξ_θ^T, ξ_z^T	Derivatives of ε_r^T, ε_θ^T and ε_z^T with respect to p
ξ_r^p, ξ_θ^p, ξ_z^p	Derivatives of ε_r^p, ε_θ^p and ε_z^T with respect to p
ρ_c	Non-dimensional radius of elastic/plastic boundary
σ_r	Radial stress
σ_θ	Circumferential stress
τ	Non-dimensional temperature introduced in Eq. (1.28)
ψ	Auxiliary function defined separately for different yield criteria in Eqs. (1.32), (1.47) and (1.53)
ψ_a	Value of ψ at the inner radius of disc
ψ_b	Value of ψ at the outer radius of disc
ψ_c	Value of ψ at elastic/plastic boundary

Chapter 1
Axisymmetric Thermo-Elastic-Plastic Problem Under Plane Stress Conditions

1.1 Introduction

Thin annular discs subject to various loading conditions are a class of commonly used structures in mechanical engineering. Particular examples are aircraft structures [1] and reciprocating machinery [2]. The mechanical analysis and design of such discs may be based either on elastic or elastic/plastic solutions. The present monograph deals with the latter approach. The Tresca yield criterion has been long associated with the solution to the stresses and strains within axisymmetric discs under plane stress conditions. Particular solutions for discs subject to various loading conditions are proposed in [3–11] where further references can be found. Another widely used simplified assumption combined with both Tresca and Mises yield criteria is the deformation theory of plasticity. Particular solutions for discs subject to various loading conditions are proposed in [7, 12–17] where further references can be found. Most of these solutions for the von Mises criterion are numerical. In some cases using deformation theories of plasticity is justified since the stress path is nearly proportional [18–21]. It is however evident that this conclusion depends on the particular boundary value problem and, possibly, the material model. A comparison of flow and deformation theories of plasticity in the case of radially stressed annular plates has been made in [22]. There are numerous numerical solutions for the von Mises yield criterion and its associated flow rule. Even though analytical and semi-analytical solutions involve more assumptions than numerical solutions, the former are convenient to compare qualitative features of boundary value problems solved for different models. Also, analytical and semi-analytical solutions are in general necessary to verify numerical codes [23, 24]. Analytical and semi-analytical solutions are usually available under plane strain or plane stress assumptions. Experimental observations indicate that the assumption of plane strain is inappropriate for thin plates and discs [25]. Also, the assumption of plane stress is often adopted even for long tubes with open ends ([26–28] among others). Moreover, it is worthy of note that numerical plane stress solutions are in general in good agreement with experiment even for plates with such strong stress and strain concentrators as cracks [29].

S. Alexandrov, *Elastic/Plastic Discs Under Plane Stress Conditions*,
SpringerBriefs in Computational Mechanics, DOI 10.1007/978-3-319-14580-8_1

Therefore, plane stress conditions are accepted in the present monograph. The usefulness of plane stress solutions in practical applications is confirmed by continued efforts to develop special purpose numerical codes [30–34].

The von Mises and Tresca yield criteria describe the initiation of yielding in isotropic pressure-independent materials. Many metallic materials are plastically anisotropic. In particular, the orthotropic form of initial anisotropy is most common, arising from such processing methods as rolling, drawing and extrusion [35]. A great number of anisotropic yield criteria have been proposed in the literature (see, [35]). The quadratic yield criterion proposed in [26] is simplest and most widely accepted. This yield criterion is adopted in the present monograph.

The yield criterion of some metallic materials depends on the hydrostatic pressure [36–41]. According to [37, 40, 41] a suitable pressure-dependent yield criterion for a wide class of metallic materials is the one proposed in [42]. All pressure dependent solutions considered in the present monograph are based on this yield criterion. In contrast to classical plasticity there is no commonly accepted relation to connect stress and strain or strain rate for pressure sensitive materials. Two assumptions are used in the present monograph. One of these assumptions is the associated flow rule. In this case the material is plastically compressible. According to the other assumption the plastic potential is given by the von Mises function. In this case the material is plastically incompressible.

Some of the solutions presented in this monograph have been proposed in [43–48]. The method developed and used in these papers has been extended to other loading conditions and materials with temperature dependent properties in [49–55].

1.2 Basic Equations

The present monograph deals with axisymmetric elastic/plastic problems under plane stress conditions. Therefore, the most general configuration that can be considered is a thin hollow disc of variable thickness. It is however assumed that the thickness is constant. For this reason the thickness is not involved in the basic equations and solutions. The inner radius of the disc is denoted by a_0 and the outer radius by b_0. Plates with a hole can be considered as a special case when $b_0 \to \infty$. It is possible to introduce a cylindrical coordinate system $(r,\ \theta,\ z)$ such that the normal stresses in this coordinate system are the principal stresses (Fig. 1.1). These stresses are denoted by σ_r, σ_θ and σ_z. It is evident that $\sigma_z = 0$ under plane stress conditions. Moreover, all the solutions presented are independent of θ and z. The classical Duhamel-Neumann law is adopted. In particular, the elastic portions of the total strains, ε_r^e, ε_θ^e and ε_z^e, are related to the stresses as

$$\varepsilon_r^e = \frac{\sigma_r - \nu\sigma_\theta}{E}, \quad \varepsilon_\theta^e = \frac{\sigma_\theta - \nu\sigma_r}{E}, \quad \varepsilon_z^e = -\frac{\nu\left(\sigma_r + \sigma_\theta\right)}{E} \tag{1.1}$$

Fig. 1.1 Geometry of disc and coordinate system

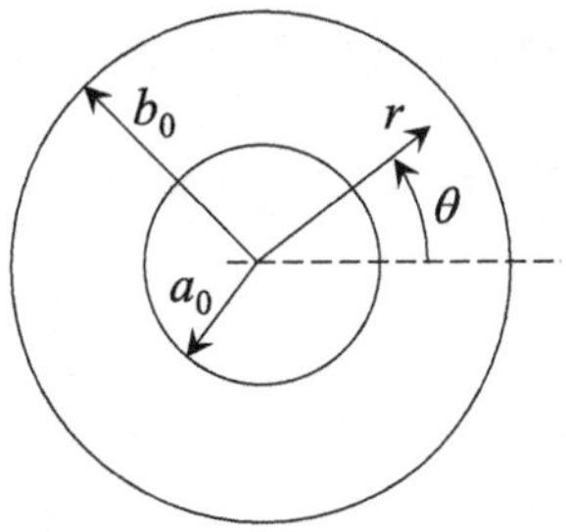

where E is Young's modulus and ν is Poisson's ratio. The thermal portions of the total strains, ε_r^T, ε_θ^T and ε_z^T, are given by

$$\varepsilon_r^T = \varepsilon_\theta^T = \varepsilon_z^T = \gamma T \tag{1.2}$$

where T the rise of temperature above the reference state and γ is the thermal coefficient of linear expansion. The total strains in plastic regions are

$$\varepsilon_r = \varepsilon_r^T + \varepsilon_r^e + \varepsilon_r^p, \quad \varepsilon_\theta = \varepsilon_\theta^T + \varepsilon_\theta^e + \varepsilon_\theta^p, \quad \varepsilon_z = \varepsilon_z^T + \varepsilon_z^e + \varepsilon_z^p \tag{1.3}$$

where ε_r^p, ε_θ^p and ε_z^p are the plastic portions of the total strains.

In what follows, several yield criteria are used. In particular, the von Mises yield criterion is adopted for isotropic pressure—independent plastic material. In terms of the principal stresses this criterion can be written as

$$(\sigma_r - \sigma_\theta)^2 + (\sigma_\theta - \sigma_z)^2 + (\sigma_z - \sigma_r)^2 = 2\sigma_Y^2 \tag{1.4}$$

where σ_Y is the yield stress in tension, a material constant. Under plane stress conditions Eq. (1.4) reduces to

$$\sigma_r^2 + \sigma_\theta^2 - \sigma_\theta\sigma_r = \sigma_Y^2. \tag{1.5}$$

In order to account for plastic anisotropy, the extension of the von Mises yield criterion proposed in [26] is used. It is assumed that the principal axes of anisotropy and the base vectors of the cylindrical system of coordinates coincide everywhere. Under this assumption the criterion [26] can be written as

$$H\,(\sigma_r - \sigma_\theta)^2 + F\,(\sigma_\theta - \sigma_z)^2 + G\,(\sigma_z - \sigma_r)^2 = 1 \tag{1.6}$$

where G, H and F are parameters which characterize the current state of anisotropy. These parameters are expressible in terms of the yield stresses in respect of the principal axes of anisotropy. Let X, Y and Z be the yield stresses in the $r-$, $\theta-$ and $z-$ directions, respectively. Then,

$$2F = \frac{1}{Y^2} + \frac{1}{Z^2} - \frac{1}{X^2}, \quad 2G = \frac{1}{Z^2} + \frac{1}{X^2} - \frac{1}{Y^2}, \quad 2H = \frac{1}{X^2} + \frac{1}{Y^2} - \frac{1}{Z^2}. \tag{1.7}$$

Under plane stress conditions Eq. (1.6) reduces to

$$(G + H)\sigma_r^2 - 2H\sigma_r\sigma_\theta + (H + F)\sigma_\theta^2 = 1.$$

It is convenient to rewrite this equation as [56]

$$\sigma_r^2 + p_\theta^2 - \eta\sigma_r p_\theta = \sigma_A^2 \tag{1.8}$$

where

$$p_\theta = \frac{\sigma_\theta}{\eta_1}, \quad \eta = 2HXY, \quad \eta_1 = \frac{Y}{X}, \quad \sigma_A = \frac{1}{G + H}. \tag{1.9}$$

In the case of isotropic materials, $X = Y = Z = \sigma_Y$. Then, it follows from Eq. (1.7) that $F = G = H$ and from Eq. (1.9) that $\eta = \eta_1 = 1$ and $\sigma_A = \sigma_Y$. At these values of η, η_1 and σ_A the yield criterion (1.8) coincides with the yield criterion (1.5).

Plastic yielding of some metallic materials is affected by the hydrostatic stress [36–41]. In many cases the yield criterion proposed in [42] is adequate for such materials. In terms of the principal stresses this criterion can be written as

$$\frac{\alpha}{3}(\sigma_r + \sigma_\theta + \sigma_z) + \frac{1}{\sqrt{2}}\sqrt{(\sigma_r - \sigma_\theta)^2 + (\sigma_\theta - \sigma_z)^2 + (\sigma_z - \sigma_r)^2} = \sigma_S \tag{1.10}$$

where α and σ_S are material constants. Under plane stress conditions this equation reduces to

$$\left(1 - \frac{\alpha^2}{9}\right)\sigma_r^2 + \left(1 - \frac{\alpha^2}{9}\right)\sigma_\theta^2 - \left(1 + \frac{2\alpha^2}{9}\right)\sigma_r\sigma_\theta + \\ + \frac{2\alpha}{3}\sigma_S(\sigma_r + \sigma_\theta) = \sigma_S^2. \tag{1.11}$$

This equation is valid if $\alpha(\sigma_r + \sigma_\theta) \leq 3\sigma_S$. It is evident that the yield criterion (1.11) reduces to the yield criterion (1.5) at $\alpha = 0$ and $\sigma_S = \sigma_Y$.

The associated flow rule is commonly accepted to connect stresses and strain rates in pressure-independent plasticity. Applying this rule to the yield criterion (1.4) results in

$$\dot{\varepsilon}_r^p = \lambda_1(2\sigma_r - \sigma_\theta - \sigma_z), \quad \dot{\varepsilon}_\theta^p = \lambda_1(2\sigma_\theta - \sigma_r - \sigma_z), \\ \dot{\varepsilon}_z^p = \lambda_1(2\sigma_z - \sigma_\theta - \sigma_r) \tag{1.12}$$

where $\dot{\varepsilon}_r^p$, $\dot{\varepsilon}_\theta^p$ and $\dot{\varepsilon}_z^p$ are the plastic portions of the total strain rates and λ_1 is a non-negative scalar factor. Under plane stress conditions Eq. (1.12) reduce to

$$\dot{\varepsilon}_r^p = \lambda_1 (2\sigma_r - \sigma_\theta), \quad \dot{\varepsilon}_\theta^p = \lambda_1 (2\sigma_\theta - \sigma_r), \quad \dot{\varepsilon}_z^p = -\lambda_1 (\sigma_\theta + \sigma_r). \tag{1.13}$$

Applying the associated flow rule to the yield criterion (1.6) results in

$$\dot{\varepsilon}_r^p = \lambda_1 \left[(H+G)\sigma_r - H\sigma_\theta - G\sigma_z\right], \quad \dot{\varepsilon}_\theta^p = \lambda_1 \left[(H+F)\sigma_\theta - H\sigma_r - F\sigma_z\right],$$
$$\dot{\varepsilon}_z^p = \lambda_1 \left[(G+F)\sigma_z - G\sigma_r - F\sigma_\theta\right].$$

Under plane stress conditions these equations reduce to

$$\dot{\varepsilon}_r^p = \lambda_1 (H+G)\left(\sigma_r - \frac{\eta}{2} p_\theta\right), \quad \dot{\varepsilon}_\theta^p = \lambda_1 \frac{(H+G)}{\eta_1}\left(p_\theta - \frac{\eta}{2}\sigma_r\right), \tag{1.14}$$
$$\dot{\varepsilon}_z^p = -\lambda_1 (H+G)\left[\left(1 - \frac{\eta}{2\eta_1}\right)\sigma_r + \left(\frac{1}{\eta_1} - \frac{\eta}{2}\right)p_\theta\right].$$

Here Eq. (1.9) has been taken into account. There is no commonly accepted equation that connects stresses and strain rates in pressure-dependent plasticity. The present monograph focuses on two widely used assumptions. One of these assumptions is that the plastic potential is given by Eq. (1.4). In this case Eq. (1.13) are valid and the material is plastically incompressible. The other assumption is the associated flow rule. Applying this rule to the yield criterion (1.10) results in

$$\dot{\varepsilon}_r^p = \lambda_2 \left[\frac{\alpha}{3} + \frac{2\sigma_r - \sigma_\theta - \sigma_z}{\sqrt{2}\sqrt{(\sigma_r - \sigma_\theta)^2 + (\sigma_\theta - \sigma_z)^2 + (\sigma_z - \sigma_r)^2}}\right],$$
$$\dot{\varepsilon}_\theta^p = \lambda_2 \left[\frac{\alpha}{3} + \frac{2\sigma_\theta - \sigma_r - \sigma_z}{\sqrt{2}\sqrt{(\sigma_r - \sigma_\theta)^2 + (\sigma_\theta - \sigma_z)^2 + (\sigma_z - \sigma_r)^2}}\right],$$
$$\dot{\varepsilon}_z^p = \lambda_2 \left[\frac{\alpha}{3} + \frac{2\sigma_z - \sigma_\theta - \sigma_r}{\sqrt{2}\sqrt{(\sigma_r - \sigma_\theta)^2 + (\sigma_\theta - \sigma_z)^2 + (\sigma_z - \sigma_r)^2}}\right].$$

Substituting here $\sqrt{(\sigma_r - \sigma_\theta)^2 + (\sigma_\theta - \sigma_z)^2 + (\sigma_z - \sigma_r)^2}$ from Eq. (1.10) and taking into account that $\sigma_z = 0$ lead to

$$\dot{\varepsilon}_r^p = \lambda_1 \left[6\alpha\sigma_s + 2\left(9 - \alpha^2\right)\sigma_r - \left(2\alpha^2 + 9\right)\sigma_\theta\right],$$
$$\dot{\varepsilon}_\theta^p = \lambda_1 \left[6\alpha\sigma_s + 2\left(9 - \alpha^2\right)\sigma_\theta - \left(2\alpha^2 + 9\right)\sigma_r\right], \tag{1.15}$$
$$\dot{\varepsilon}_z^p = \lambda_1 \left[6\alpha\sigma_S - \left(9 + 2\alpha^2\right)(\sigma_r + \sigma_\theta)\right],$$
$$\lambda_1 = \frac{\lambda_2}{6\left[3\sigma_s - \alpha(\sigma_r + \sigma_\theta)\right]}.$$

It follows from these equations that in general $\dot{\varepsilon}_r^p + \dot{\varepsilon}_\theta^p + \dot{\varepsilon}_z^p \neq 0$. Therefore, the material is plastically compressible. Since strains are supposed to be small, the change in relative density caused by volumetric strain is ignored.

The models considered are rate-independent. Therefore, strain rates can be replaced with the derivatives of strains with respect to any monotonically increasing parameter p. Denote

$$\begin{aligned}
\xi_r &= \frac{\partial \varepsilon_r}{\partial p}, \quad \xi_\theta = \frac{\partial \varepsilon_\theta}{\partial p}, \quad \xi_z = \frac{\partial \varepsilon_z}{\partial p}, \\
\xi_r^T &= \frac{\partial \varepsilon_r^T}{\partial p}, \quad \xi_\theta^T = \frac{\partial \varepsilon_\theta^T}{\partial p}, \quad \xi_z^T = \frac{\partial \varepsilon_z^T}{\partial p}, \\
\xi_r^e &= \frac{\partial \varepsilon_r^e}{\partial p}, \quad \xi_\theta^e = \frac{\partial \varepsilon_\theta^e}{\partial p}, \quad \xi_z^e = \frac{\partial \varepsilon_z^e}{\partial p}, \\
\xi_r^p &= \frac{\partial \varepsilon_r^p}{\partial p}, \quad \xi_\theta^p = \frac{\partial \varepsilon_\theta^p}{\partial p}, \quad \xi_z^p = \frac{\partial \varepsilon_z^p}{\partial p}.
\end{aligned} \tag{1.16}$$

Then, Eqs. (1.13), (1.14) and (1.15) become

$$\xi_r^p = \lambda\,(2\sigma_r - \sigma_\theta)\,, \quad \xi_\theta^p = \lambda\,(2\sigma_\theta - \sigma_r)\,, \quad \xi_z^p = -\lambda\,(\sigma_\theta + \sigma_r)\,, \tag{1.17}$$

$$\begin{aligned}
\xi_r^p &= \lambda\left(\sigma_r - \frac{\eta}{2} p_\theta\right), \quad \xi_\theta^p = \frac{\lambda}{\eta_1}\left(p_\theta - \frac{\eta}{2}\sigma_r\right), \\
\xi_z^p &= -\lambda\left[\left(1 - \frac{\eta}{2\eta_1}\right)\sigma_r + \left(\frac{1}{\eta_1} - \frac{\eta}{2}\right) p_\theta\right].
\end{aligned} \tag{1.18}$$

$$\begin{aligned}
\xi_r^p &= \lambda\left[6\alpha\sigma_s + 2\left(9 - \alpha^2\right)\sigma_r - \left(2\alpha^2 + 9\right)\sigma_\theta\right], \\
\xi_\theta^p &= \lambda\left[6\alpha\sigma_s + 2\left(9 - \alpha^2\right)\sigma_\theta - \left(2\alpha^2 + 9\right)\sigma_r\right], \\
\xi_z^p &= \lambda\left[6\alpha\sigma_S - \left(9 + 2\alpha^2\right)(\sigma_r + \sigma_\theta)\right],
\end{aligned} \tag{1.19}$$

respectively. It is evident that $\lambda = \lambda_1 dp/dt$ in Eq. (1.17), $\lambda = \lambda_1 (H + G)\, dp/dt$ in Eq. (1.18) and $\lambda = \lambda_1 dp/dt$ in Eq. (1.19). In either case $\lambda \geq 0$. It is worthy of note that it is possible to assume that p is a monotonically decreasing parameter. In this case however $\lambda \leq 0$. It follows from Eqs. (1.1), (1.2), (1.3) and (1.16) that

$$\begin{aligned}
\xi_r^e &= \frac{1}{E}\left(\frac{\partial \sigma_r}{\partial p} - \nu \frac{\partial \sigma_\theta}{\partial p}\right), \quad \xi_\theta^e = \frac{1}{E}\left(\frac{\partial \sigma_\theta}{\partial p} - \nu \frac{\partial \sigma_r}{\partial p}\right), \\
\xi_z^e &= -\frac{\nu}{E}\left(\frac{\partial \sigma_r}{\partial p} + \frac{\partial \sigma_\theta}{\partial p}\right),
\end{aligned} \tag{1.20}$$

$$\xi_r^T = \xi_\theta^T = \xi_z^T = \gamma \frac{dT}{dp}, \tag{1.21}$$

$$\xi_r = \xi_r^T + \xi_r^e + \xi_r^p, \quad \xi_\theta = \xi_\theta^T + \xi_\theta^e + \xi_\theta^p, \quad \xi_z = \xi_z^T + \xi_z^e + \xi_z^p. \tag{1.22}$$

The only non-trivial equilibrium equation for the disc of uniform thickness is

$$\frac{\partial \sigma_r}{\partial r} + \frac{\sigma_r - \sigma_\theta}{r} = 0. \tag{1.23}$$

The radial and circumferential strains are expressed in terms of the radial displacement as

$$\varepsilon_r = \frac{\partial u}{\partial r}, \quad \varepsilon_\theta = \frac{u}{r}. \tag{1.24}$$

It is evident from these relations that the equations of strain and strain rate compatibility are

$$r \frac{\partial \varepsilon_\theta}{\partial r} + \varepsilon_\theta - \varepsilon_r = 0, \quad r \frac{\partial \dot{\varepsilon}_\theta}{\partial r} + \dot{\varepsilon}_\theta - \dot{\varepsilon}_r = 0.$$

The latter is equivalent to

$$r \frac{\partial \xi_\theta}{\partial r} + \xi_\theta - \xi_r = 0. \tag{1.25}$$

In the case of elastic/plastic solutions elastic and plastic zones exist. The present monograph is restricted to solutions in which there is just one plastic zone. Axial symmetry demands that the elastic/plastic boundary is a circle, $r = r_c$. The radial displacement and velocity are continuous across this boundary. Then, it follows from Eq. (1.24) that

$$[\varepsilon_\theta] = 0 \quad \text{and} \quad [\xi_\theta] = 0 \tag{1.26}$$

at $r = r_c$. Here and in what follows [...] denotes the amount of jump in the quantity enclosed in the brackets. A requirement of equilibrium is that the radial stress is continuous across the elastic/plastic boundary. The material just on the elastic side of this boundary must satisfy the yield criterion [26]. This condition is equivalent to the requirement that the circumferential stress is continuous across the elastic/plastic boundary since $\sigma_z = 0$ everywhere. Thus both σ_r and σ_θ are continuous across the elastic/plastic boundary

$$[\sigma_r] = 0 \quad \text{and} \quad [\sigma_\theta] = 0 \tag{1.27}$$

at $r = r_c$.

1.3 General Stress Solutions

In succeeding chapters the solutions for the yield criteria (1.5), (1.8) and (1.11) are given in different sections. Therefore, to simplify the notation the symbol σ_0 will be used instead of σ_Y in the solutions for the yield criterion (1.5), instead of σ_A in the solutions for the yield criterion (1.8) and instead of σ_S in the solutions for the yield criterion (1.11). It is also convenient to introduce the following dimensionless quantities

$$\rho = \frac{r}{b_0}, \quad \rho_c = \frac{r_c}{b_0}, \quad a = \frac{a_0}{b_0}, \quad k = \frac{\sigma_0}{E}, \quad \tau = \frac{\gamma T E}{\sigma_0}. \tag{1.28}$$

1.3.1 General Thermo/Elastic Solution

The general thermo/elastic solution must satisfy Eqs. (1.1)–(1.3), (1.23) and (1.24) at $\varepsilon_r^p = \varepsilon_\theta^p = \varepsilon_z^p = 0$. This solution is well known (see, for example, [26]). In particular, the solution for stresses can be written as

$$\frac{\sigma_r}{\sigma_0} = \frac{A}{\rho^2} + B, \qquad \frac{\sigma_\theta}{\sigma_0} = -\frac{A}{\rho^2} + B \tag{1.29}$$

where A and B are constants of integration. It is assumed that the disc has no stress at the initial instant. Loading applied to the disc affects the zero-stress state. The range of validity of this purely elastic solution is determined by substituting Eq. (1.29) into this or that yield criterion. In particular, the corresponding value of the loading parameter and the site of plastic zone initiation are found.

Equation (1.29) is also valid in the elastic region of elastic/plastic discs.

1.3.2 Yield Criterion (1.5)

It follows from Eqs. (1.5) and (1.29) that the range of validity of the purely elastic solution is restricted by the inequality

$$B^2 + \frac{3A^2}{\rho^4} \le 1. \tag{1.30}$$

It is worthy of note that any yield criterion defines the limit of elasticity and the elastic solution is valid as long as this limit is not reached. Therefore, the inequality sign appears in Eq. (1.30). The left hand side of Eq. (1.30) is a monotonically decreasing function of ρ. Therefore, a plastic zone starts to develop at $\rho = a$. In this case, it follows from Eq. (1.30) that

$$B^2 + \frac{3A^2}{a^4} = 1 \tag{1.31}$$

at the instant of the initiation of plastic yielding. The state of stress in the plastic zone is determined from Eqs. (1.5) and (1.23). Equation (1.5) is automatically satisfied by

$$\frac{\sigma_r}{\sigma_0} = -\frac{2\sin\psi}{\sqrt{3}}, \quad \frac{\sigma_\theta}{\sigma_0} = -\frac{\sin\psi}{\sqrt{3}} - \cos\psi \tag{1.32}$$

where ψ is a new function of ρ and a time-like variable. Substituting Eq. (1.32) into Eq. (1.23) gives

$$2\rho\cos\psi\frac{\partial\psi}{\partial\rho} = \sqrt{3}\cos\psi - \sin\psi. \tag{1.33}$$

The solution to this equation can be written as

$$\ln\frac{\rho}{\rho_0} = 2\int\limits_{\psi_0}^{\psi}\frac{\cos\mu}{\left(\sqrt{3}\cos\mu - \sin\mu\right)}d\mu. \tag{1.34}$$

Here ρ_0 is arbitrary constant but ψ_0 may depend on a time-like variable. Also, μ is a dummy variable of integration. The integral in Eq. (1.34) can be evaluated in terms of elementary functions. As a result,

$$\rho = \rho_0 \exp\left[\frac{\sqrt{3}}{2}(\psi - \psi_0)\right]\sqrt{\frac{\sin(\psi_0 - \pi/3)}{\sin(\psi - \pi/3)}}. \tag{1.35}$$

It is evident that this solution satisfies the boundary condition $\psi = \psi_0$ for $\rho = \rho_0$.

1.3.3 Yield Criterion (1.8)

It follows from Eqs. (1.8) and (1.29) that the range of validity of the purely elastic solution is restricted by the inequality

$$B^2[1 + \eta_1(\eta_1 - \eta)] + \frac{2AB\left(\eta_1^2 - 1\right)}{\rho^2} + \frac{A^2[1 + \eta_1(\eta_1 + \eta)]}{\rho^4} \le \eta_1^2. \tag{1.36}$$

This inequality can be rewritten in the form

$$g(\rho) = b_1 + \frac{b_2}{\rho^2} + \frac{b_3}{\rho^4} \le 0 \tag{1.37}$$

where

$$b_1 = B^2 [1 + \eta_1 (\eta_1 - \eta)] - \eta_1^2, \quad b_2 = 2AB \left(\eta_1^2 - 1\right), \tag{1.38}$$
$$b_3 = A^2 [1 + \eta_1 (\eta_1 + \eta)] .$$

The function $g(\rho)$ may attain a maximum at a point within the range $a < \rho < 1$ if and only if $dg/d\rho = 0$ at that point. It follows from Eq. (1.37) that

$$\frac{dg}{d\rho} = -\frac{2}{\rho^3}\left(b_2 + \frac{2b_3}{\rho^2}\right). \tag{1.39}$$

Therefore, $dg/d\rho = 0$ at

$$\rho = \rho_n = \sqrt{-\frac{2b_3}{b_2}}. \tag{1.40}$$

It is seen from Eq. (1.38) that $b_3 > 0$. Therefore, ρ_n is an imaginary number if $b_2 > 0$. Assume that $b_2 < 0$. Then, ρ_n is a real number. The function $g(\rho)$ attains a maximum at $\rho = \rho_n$ if $d^2g/d\rho^2 < 0$ at that point. Differentiating Eq. (1.39) and putting $\rho = \rho_n$ yield

$$\frac{d^2g}{d\rho^2} = \frac{8b_3}{\rho_n^6} \tag{1.41}$$

at $\rho = \rho_n$. Since $b_3 > 0$, it follows from Eq. (1.41) that $d^2g/d\rho^2 > 0$ at $\rho = \rho_n$. Therefore, the function $g(\rho)$ attains a maximum within the range $a \le \rho \le 1$ at $\rho = a$ or $\rho = 1$ independently of the value of b_2. If $b_2 > 0$ then $dg/d\rho < 0$ everywhere as follows from Eq. (1.39). In this case the function $g(\rho)$ attains its maximum in the range $a \le \rho \le 1$ at $\rho = a$ and a plastic zone starts to develop from this radius. The corresponding condition is determined from Eq. (1.37) as

$$b_1 + \frac{b_2}{a^2} + \frac{b_3}{a^4} = 0. \tag{1.42}$$

If $b_2 < 0$ then it is necessary to consider the expression

$$E = g(1) - g(a) = \left(1 - \frac{1}{a^2}\right)\left[b_2 + b_3\left(1 + \frac{1}{a^2}\right)\right]. \tag{1.43}$$

It is evident that a plastic zone starts to develop from the inner radius if $E < 0$ and from the outer radius if $E > 0$. Two plastic zones start to develop from the inner and outer radii simultaneously if $E = 0$. Since $a < 1$, it follows from Eq. (1.43) that the condition $E < 0$ is equivalent to

$$b_2 + b_3\left(1 + \frac{1}{a^2}\right) > 0, \tag{1.44}$$

the condition $E > 0$ to

$$b_2 + b_3\left(1 + \frac{1}{a^2}\right) < 0, \tag{1.45}$$

and the condition $E = 0$ to

$$b_2 + b_3\left(1 + \frac{1}{a^2}\right) = 0. \tag{1.46}$$

The state of stress in plastic zones is determined from Eqs. (1.8) and (1.23). Equation (1.8) is automatically satisfied by

$$\frac{\sigma_r}{\sigma_0} = -\frac{2\sin\psi}{\sqrt{4-\eta^2}}, \quad \frac{p_\theta}{\sigma_0} = -\frac{\eta\sin\psi}{\sqrt{4-\eta^2}} - \cos\psi \tag{1.47}$$

where ψ is a new function of ρ and a time-like variable. Substituting this equation into Eq. (1.23) and using Eq. (1.9) to replace p_θ with σ_θ give

$$2\rho\frac{\partial\psi}{\partial\rho} = \eta_1\sqrt{4-\eta^2} - (2-\eta\eta_1)\tan\psi. \tag{1.48}$$

The solution to this equation can be written as

$$\ln\frac{\rho}{\rho_0} = 2\int\limits_{\psi_0}^{\psi}\frac{d\mu}{\left[\eta_1\sqrt{4-\eta^2} - (2-\eta\eta_1)\tan\mu\right]}. \tag{1.49}$$

Here ρ_0 is arbitrary constant but ψ_0 may depend on a time-like variable. The integral in Eq. (1.49) can be evaluated in terms of elementary functions. As a result,

$$\begin{aligned}\ln\left(\frac{\rho}{\rho_0}\right) &= \frac{\eta_1\sqrt{4-\eta^2}}{2\left[1+\eta_1(\eta_1-\eta)\right]}(\psi-\psi_0) + \\ &+ \frac{(2-\eta\eta_1)}{2\left[1+\eta_1(\eta_1-\eta)\right]}\ln\left[\frac{\eta_1\sqrt{4-\eta^2}\cos\psi_0 - (2-\eta\eta_1)\sin\psi_0}{\eta_1\sqrt{4-\eta^2}\cos\psi - (2-\eta\eta_1)\sin\psi}\right].\end{aligned} \tag{1.50}$$

This solution reduces to Eq. (1.35) at $\eta = \eta_1 = 1$ and satisfies the condition $\psi = \psi_0$ at $\rho = \rho_0$.

1.3.4 Yield Criterion (1.11)

It follows from Eqs. (1.11) and (1.29) that the range of validity of the purely elastic solution is restricted by the inequality

$$\frac{3A^2}{\rho^4} + \left(1 - \frac{4\alpha^2}{9}\right) B^2 + \frac{4\alpha B}{3} \le 1. \tag{1.51}$$

The right hand side of Eq. (1.51) is a monotonically decreasing function of ρ. Therefore, a plastic zone starts to develop at $\rho = a$. It follows from Eq. (1.51) that

$$\frac{3A^2}{a^4} + \left(1 - \frac{4\alpha^2}{9}\right) B^2 + \frac{4\alpha B}{3} = 1. \tag{1.52}$$

at the instant of the initiation of plastic yielding. The state of stress in the plastic zone is determined from Eqs. (1.11) and (1.23). Equation (1.11) is automatically satisfied by

$$\frac{\sigma_r}{\sigma_0} = 3\beta_0 - \frac{\beta_1}{2}\left(1 + 3\sqrt{3}\beta_1\right)\sin\psi + \frac{\sqrt{3}}{2}\beta_1\left(1 - \sqrt{3}\beta_1\right)\cos\psi, \tag{1.53}$$

$$\frac{\sigma_\theta}{\sigma_0} = 3\beta_0 + \frac{\beta_1}{2}\left(1 - 3\sqrt{3}\beta_1\right)\sin\psi - \frac{\sqrt{3}}{2}\beta_1\left(1 + \sqrt{3}\beta_1\right)\cos\psi$$

where ψ is a new function of ρ and a time-like variable. Also,

$$\beta_0 = \frac{2\alpha}{4\alpha^2 - 9}, \quad \beta_1 = \frac{\sqrt{3}}{\sqrt{9 - 4\alpha^2}}. \tag{1.54}$$

Substituting Eq. (1.53) into Eq. (1.23) gives

$$\left[\left(1 + 3\sqrt{3}\beta_1\right)\cos\psi + \sqrt{3}\left(1 - \sqrt{3}\beta_1\right)\sin\psi\right]\frac{\partial\psi}{\partial\rho} = \tag{1.55}$$

$$= \frac{2\left(\sqrt{3}\cos\psi - \sin\psi\right)}{\rho}.$$

The solution to this equation can be written as

$$\ln\frac{\rho}{\rho_0} = \frac{1}{2}\int_{\psi_0}^{\psi} \frac{\left[\left(1 + 3\sqrt{3}\beta_1\right)\cos\mu + \sqrt{3}\left(1 - \sqrt{3}\beta_1\right)\sin\mu\right]}{\sqrt{3}\cos\mu - \sin\mu} d\mu. \tag{1.56}$$

Here ρ_0 is arbitrary constant but ψ_0 may depend on a time-like variable. The integral in Eq. (1.56) can be evaluated in terms of elementary functions. As a result,

$$\rho = \rho_0 \exp\left[\frac{3\beta_1}{2}(\psi - \psi_0)\right]\sqrt{\frac{\sin(\psi_0 - \pi/3)}{\sin(\psi - \pi/3)}}. \tag{1.57}$$

It is seen from Eq. (1.54) that $\beta_1 = 1/\sqrt{3}$ at $\alpha = 0$. Therefore, Eq. (1.57) reduces to Eq. (1.35) at $\alpha = 0$. It is evident that the solution (1.57) satisfies the condition $\psi = \psi_0$ at $\rho = \rho_0$.

1.4 General Strain Solutions

It follows from Eqs. (1.21) and (1.28) that

$$\xi_r^T = \xi_\theta^T = \xi_z^T = k\frac{d\tau}{dp}. \tag{1.58}$$

This equation is valid in elastic and plastic regions. Using Eqs. (1.22), (1.28) and (1.58) it is possible to rewrite Eq. (1.25) as

$$\rho\frac{\partial \xi_\theta}{\partial \rho} + \xi_\theta = \xi_r^p + \xi_r^e + k\frac{d\tau}{dp}. \tag{1.59}$$

It will be seen later that it is convenient to represent the boundary conditions (1.26) and (1.27) in terms of ψ. To this end, the value of ψ at $\rho = \rho_c$ is denoted by ψ_c. It is worthy of note that different relations are used to introduce the function ψ in the case of different yield criteria. These relations are given in Eqs. (1.32), (1.47) and (1.53). The notation ψ_c is used independently of the yield criterion.

1.4.1 General Thermo/Elastic Solution

The distribution of the elastic portion of the strain tensor associated with the stress solution (1.29) follows from Eq. (1.1) with the use of Eq. (1.28) as

$$\frac{\varepsilon_r^e}{k} = \frac{A(1+\nu)}{\rho^2} + B(1-\nu), \quad \frac{\varepsilon_\theta^e}{k} = -\frac{A(1+\nu)}{\rho^2} + B(1-\nu), \quad \frac{\varepsilon_z^e}{k} = -2\nu B. \tag{1.60}$$

The thermal portion of the strain tensor is given by Eq. (1.2). Therefore, using Eqs. (1.3), (1.28) and (1.60) the total strains are determined as

$$\frac{\varepsilon_r}{k} = \frac{A(1+\nu)}{\rho^2} + B(1-\nu) + \tau, \quad \frac{\varepsilon_\theta}{k} = -\frac{A(1+\nu)}{\rho^2} + B(1-\nu) + \tau, \tag{1.61}$$
$$\frac{\varepsilon_z}{k} = -2\nu B + \tau.$$

Equations (1.60) and (1.61) are valid in purely elastic discs and in the elastic region of elastic/plastic discs. Using Eq. (1.28) it is convenient to rewrite the condition $(1.26)^2$ as

$$\xi_\theta = \xi_c \tag{1.62}$$

for $\rho = \rho_c$ (or $\psi = \psi_c$). The value of ξ_c is determined using the solution in the elastic region. In particular, it follows from Eq. (1.61) that

$$\frac{\xi_c}{k} = -\frac{(1+\nu)}{\rho_c^2}\frac{dA}{dp} + (1-\nu)\frac{dB}{dp} + \frac{d\tau}{dp}. \tag{1.63}$$

Equations (1.62) and (1.63) provide the boundary condition for calculating ξ_θ in the plastic region. Since ξ_θ^T is independent of ρ, as follows from Eq. (1.21), the condition given in Eq. (1.26)^2 is equivalent to $\left[\xi_\theta^e + \xi_\theta^p\right] = 0$ at $\rho = \rho_c$. This condition can be rewritten as

$$\xi_\theta^e + \xi_\theta^p = \xi_c^p \tag{1.64}$$

for $\rho = \rho_c$ (or $\psi = \psi_c$). The value of ξ_c^p is determined using the solution in the elastic region. In particular, it follows from Eq. (1.60) that

$$\frac{\xi_c^p}{k} = -\frac{(1+\nu)}{\rho_c^2}\frac{dA}{dp} + (1-\nu)\frac{dB}{dp}. \tag{1.65}$$

The value of ξ_θ^e involved in Eq. (1.64) is found from Eq. (1.20) using the stress solution in the plastic region. Therefore, Eqs. (1.64) and (1.65) provide the boundary condition for calculating ξ_θ^p in the plastic region.

1.4.2 Yield Criterion (1.5)

In general, it is necessary to use Eq. (1.22) instead of Eq. (1.3) in the plastic zone. Equation (1.32) is valid in this zone. Substituting this equation into Eq. (1.20) yields

$$\frac{\xi_r^e}{k} = -\left[\frac{(2-\nu)\cos\psi}{\sqrt{3}} + \nu\sin\psi\right]\frac{\partial\psi}{\partial p}, \tag{1.66}$$
$$\frac{\xi_\theta^e}{k} = \left[\sin\psi - \frac{(1-2\nu)\cos\psi}{\sqrt{3}}\right]\frac{\partial\psi}{\partial p}, \quad \frac{\xi_z^e}{k} = -2\nu\sin\left(\psi - \frac{\pi}{3}\right)\frac{\partial\psi}{\partial p}$$

in the plastic zone. Substituting Eq. (1.32) into Eq. (1.17) leads to

$$\xi_r^p = -2\sigma_0\lambda\sin\left(\psi - \frac{\pi}{6}\right), \tag{1.67}$$
$$\xi_\theta^p = -2\sigma_0\lambda\cos\psi, \quad \xi_z^p = 2\sigma_0\lambda\sin\left(\psi + \frac{\pi}{6}\right).$$

Eliminating λ between these equations gives

$$\xi_r^p = \xi_\theta^p \frac{\sin(\psi - \pi/6)}{\cos\psi}, \quad \xi_z^p = -\xi_\theta^p \frac{\sin(\psi + \pi/6)}{\cos\psi}. \tag{1.68}$$

It is seen from Eq. (1.66) that it is necessary to find the derivative $\partial\psi/\partial p$. Differentiating Eq. (1.34) yields

$$\frac{\cos\psi d\psi}{\left(\sqrt{3}\cos\psi - \sin\psi\right)} = \frac{d\rho}{2\rho} + \frac{\cos\psi_0 d\psi_0}{\left(\sqrt{3}\cos\psi_0 - \sin\psi_0\right)}.$$

It follows from this equation that

$$\frac{\partial\psi}{\partial p} = \frac{\Omega_M(\psi_0)\,d\psi_0}{\Omega_M(\psi)\,dp}, \quad \Omega_M(x) = \frac{\cos x}{\left(\sqrt{3}\cos x - \sin x\right)}. \tag{1.69}$$

Eliminating ξ_r^p in Eq. (1.59) by means of Eq. (1.68) and, then, ξ_θ^p by means of Eqs. (1.22) and (1.58) lead to

$$2\rho\frac{\partial\xi_\theta}{\partial\rho} + \sqrt{3}\left(\sqrt{3} - \tan\psi\right)\xi_\theta = \left(1 - \sqrt{3}\tan\psi\right)\xi_\theta^e + 2\xi_r^e + \sqrt{3}k\left(\sqrt{3} - \tan\psi\right)\frac{d\tau}{dp}.$$

Replacing differentiation with respect to ρ with differentiation with respect to ψ by means of Eq. (1.33) and eliminating ξ_r^e and ξ_θ^e by means of Eqs. (1.66) and (1.69) result in

$$\frac{\partial\xi_\theta}{\partial\psi} + \sqrt{3}\xi_\theta = \tag{1.70}$$
$$= \frac{k\Omega_M(\psi_0)\,d\psi_0}{\sqrt{3}\quad dp}\frac{\left[(1-2\nu)\left(\sqrt{3}\sin 2\psi - \cos 2\psi\right) - 2(2-\nu)\right]}{\cos\psi} + \sqrt{3}k\frac{d\tau}{dp}.$$

This is a linear ordinary differential equation for ξ_θ. Therefore, its general solution can be found with no difficulty. The solution to Eq. (1.70) satisfying the boundary condition (1.62) is

$$\frac{\xi_\theta}{k} = \frac{\xi_c}{k}\exp\left[\sqrt{3}(\psi_c - \psi)\right] + \frac{d\tau}{dp}\left\{1 - \exp\left[\sqrt{3}(\psi_c - \psi)\right]\right\} + \tag{1.71}$$
$$+ \frac{\Omega(\psi_0)\,d\psi_0}{\sqrt{3}\quad dp}\int\limits_{\psi_c}^{\psi}\frac{\left[(1-2\nu)\left(\sqrt{3}\sin 2\mu - \cos 2\mu\right) - 2(2-\nu)\right]}{\cos\mu}\exp\left[\sqrt{3}(\mu - \psi)\right]d\mu.$$

where ξ_c is given by Eq. (1.63). Using Eqs. (1.66), (1.69) and (1.71) it is possible to find ξ_θ^p from Eqs. (1.22) and (1.58) as $\xi_\theta^p = \xi_\theta - \xi_\theta^e - kd\tau/dp$. Then, ξ_r^p and ξ_z^p are determined from Eq. (1.68). Equations (1.22), (1.58), (1.66) and (1.69) supply ξ_r and ξ_z as functions of ψ and p. Since ψ depends on ρ and p, the equations for calculating the strains are

$$\frac{\partial \varepsilon_r}{\partial p} + \frac{\partial \varepsilon_r}{\partial \psi}\frac{\partial \psi}{\partial p} = \xi_r, \quad \frac{\partial \varepsilon_\theta}{\partial p} + \frac{\partial \varepsilon_\theta}{\partial \psi}\frac{\partial \psi}{\partial p} = \xi_\theta, \quad \frac{\partial \varepsilon_z}{\partial p} + \frac{\partial \varepsilon_z}{\partial \psi}\frac{\partial \psi}{\partial p} = \xi_z. \tag{1.72}$$

Using (1.69) to eliminate the derivative $\partial \psi / \partial p$ gives

$$\frac{\partial \varepsilon_r}{\partial p} + \frac{\partial \varepsilon_r}{\partial \psi}\frac{\Omega_M(\psi_0)\, d\psi_0}{\Omega_M(\psi)\, dp} = \xi_r, \quad \frac{\partial \varepsilon_\theta}{\partial p} + \frac{\partial \varepsilon_\theta}{\partial \psi}\frac{\Omega_M(\psi_0)\, d\psi_0}{\Omega_M(\psi)\, dp} = \xi_\theta, \quad \frac{\partial \varepsilon_z}{\partial p} + \frac{\partial \varepsilon_z}{\partial \psi}\frac{\Omega_M(\psi_0)\, d\psi_0}{\Omega_M(\psi)\, dp} = \xi_z.$$

Each of these equations is equivalent to a system of ordinary differential equations. In particular,

$$\Omega_M(\psi_0)\, d\psi_0 = \Omega_M(\psi)\, d\psi = \frac{d\psi_0}{dp}\frac{\Omega_M(\psi_0)\, d\varepsilon_r}{\xi_r},$$
$$\Omega_M(\psi_0)\, d\psi_0 = \Omega_M(\psi)\, d\psi = \frac{d\psi_0}{dp}\frac{\Omega_M(\psi_0)\, d\varepsilon_\theta}{\xi_\theta}, \tag{1.73}$$
$$\Omega_M(\psi_0)\, d\psi_0 = \Omega_M(\psi)\, d\psi = \frac{d\psi_0}{dp}\frac{\Omega_M(\psi_0)\, d\varepsilon_z}{\xi_z}.$$

Therefore, the characteristic equation is

$$\frac{d\psi}{d\psi_0} = \frac{\Omega_M(\psi_0)}{\Omega_M(\psi)} \tag{1.74}$$

and the compatibility equations are

$$\frac{d\varepsilon_r}{dp} = \xi_r, \quad \frac{d\varepsilon_\theta}{dp} = \xi_\theta, \quad \frac{d\varepsilon_z}{dp} = \xi_z. \tag{1.75}$$

It is understood here that ψ involved in ξ_r, ξ_θ and ξ_z is eliminated by means of the solution to Eq. (1.74).

1.4.3 Yield Criterion (1.8)

Substituting Eq. (1.47) in which p_θ should be eliminated by means of Eq. (1.9) into Eq. (1.20) yields

$$\frac{\xi_r^e}{k} = \left[\frac{(\nu\eta\eta_1 - 2)\cos\psi}{\sqrt{4-\eta^2}} - \nu\eta_1\sin\psi\right]\frac{\partial\psi}{\partial p},$$
$$\frac{\xi_\theta^e}{k} = \left[\eta_1\sin\psi + \frac{(2\nu - \eta\eta_1)\cos\psi}{\sqrt{4-\eta^2}}\right]\frac{\partial\psi}{\partial p}, \tag{1.76}$$
$$\frac{\xi_z^e}{k} = \nu\left[\frac{(2+\eta\eta_1)}{\sqrt{4-\eta^2}}\cos\psi - \eta_1\sin\psi\right]\frac{\partial\psi}{\partial p}$$

in the plastic zone. Substituting Eq. (1.47) into Eq. (1.18) leads to

$$\xi_r^p = \frac{\sigma_0\lambda}{2}\left(\eta\cos\psi - \sqrt{4-\eta^2}\sin\psi\right), \quad \xi_\theta^p = -\frac{\sigma_0\lambda}{\eta_1}\cos\psi, \tag{1.77}$$
$$\xi_z^p = \frac{\sigma_0\lambda}{2}\left[\left(\frac{2}{\eta_1} - \eta\right)\cos\psi + \sqrt{4-\eta^2}\sin\psi\right].$$

Eliminating λ between these equations gives

$$\xi_r^p = \frac{\xi_\theta^p\eta_1}{2}\left(\sqrt{4-\eta^2}\tan\psi - \eta\right), \tag{1.78}$$
$$\xi_z^p = -\frac{\xi_\theta^p}{2}\left(2 - \eta\eta_1 + \eta_1\sqrt{4-\eta^2}\tan\psi\right).$$

It is seen from Eq. (1.76) that it is necessary to find the derivative $\partial\psi/\partial p$. Differentiating Eq. (1.49) yields

$$\frac{d\psi}{\eta_1\sqrt{4-\eta^2} - (2-\eta\eta_1)\tan\psi} = \frac{d\psi_0}{\eta_1\sqrt{4-\eta^2} - (2-\eta\eta_1)\tan\psi_0} + \frac{d\rho}{2\rho}.$$

It follows from this equation that

$$\frac{\partial\psi}{\partial p} = \frac{\Omega_A(\psi_0)}{\Omega_A(\psi)}\frac{d\psi_0}{dp}, \quad \Omega_A(x) = \left[\eta_1\sqrt{4-\eta^2} - (2-\eta\eta_1)\tan x\right]^{-1}. \tag{1.79}$$

It is seen from Eqs. (1.69) and (1.79) that $\Omega_A(x) \equiv \Omega_M(x)$ if $\eta = \eta_1 = 1$. Eliminating ξ_r^p in Eq. (1.59) by means of Eq. (1.78) and, then, ξ_θ^p by means of Eqs. (1.22) and (1.58) lead to

$$2\rho\frac{\partial\xi_\theta}{\partial\rho}+\xi_\theta\left[2-\eta_1\left(\sqrt{4-\eta^2}\tan\psi-\eta\right)\right]=$$
$$=2\xi_r^e-\eta_1\left(\sqrt{4-\eta^2}\tan\psi-\eta\right)\xi_\theta^e+k\left[2-\eta_1\left(\sqrt{4-\eta^2}\tan\psi-\eta\right)\right]\frac{d\tau}{dp}.$$

Replacing here differentiation with respect to ρ with differentiation with respect to ψ by means of Eq. (1.48) and eliminating ξ_r^e and ξ_θ^e by means of Eqs. (1.76) and (1.79) result in

$$\frac{\partial\xi_\theta}{\partial\psi}=W_0(\psi)\,\xi_\theta+k\Omega_A(\psi_0)\frac{d\psi_0}{dp}W_1(\psi)-k\frac{d\tau}{dp}W_0(\psi)\tag{1.80}$$

where

$$W_0(\psi)=\frac{\sqrt{4-\eta^2}\eta_1\tan\psi-\eta\eta_1-2}{\sqrt{4-\eta^2}\eta_1+(\eta\eta_1-2)\tan\psi},$$
$$W_1(\psi)=\eta_1\sin\psi\left[2(\eta\eta_1-2\nu)-\sqrt{4-\eta^2}\eta_1\tan\psi\right]+\tag{1.81}$$
$$+\frac{[\eta\eta_1(4\nu-\eta\eta_1)-4]\cos\psi}{\sqrt{4-\eta^2}}.$$

Equation (1.80) is a linear ordinary differential equation for ξ_θ. The solution to this equation satisfying the boundary condition (1.62) is

$$\frac{\xi_\theta}{k}=\left\{\begin{array}{l}\Omega_A(\psi_0)\frac{d\psi_0}{dp}\int\limits_{\psi_c}^{\psi}\exp\left[-\int\limits_{\psi_c}^{\mu_1}W_0(\mu)\,d\mu\right]W_1(\mu_1)\,d\mu_1-\\-\frac{d\tau}{dp}\int\limits_{\psi_c}^{\psi}\exp\left[-\int\limits_{\psi_c}^{\mu_1}W_0(\mu)\,d\mu\right]W_0(\mu_1)\,d\mu_1+\frac{\xi_c}{k}\end{array}\right\}\times\tag{1.82}$$
$$\times\exp\left[\int\limits_{\psi_c}^{\psi}W_0(\mu)\,d\mu\right].$$

Here both μ and μ_1 are dummy variables of integration. The value of ξ_c is given by Eq. (1.63). Using Eqs. (1.76), (1.79) and (1.82) it is possible to find ξ_θ^p from Eqs. (1.22) and (1.58) as $\xi_\theta^p=\xi_\theta-\xi_\theta^e-kd\tau/dp$. Then, ξ_r^p and ξ_z^p are determined from Eq. (1.78). Equations (1.22), (1.58), (1.76) and (1.79) supply ξ_r and ξ_z as functions of ψ and p. Equation (1.72) are valid. Eliminating the derivative $\partial\psi/\partial p$ by means of Eq. (1.79) leads to the characteristic equation in the form

$$\frac{d\psi}{d\psi_0}=\frac{\Omega_A(\psi_0)}{\Omega_A(\psi)}\tag{1.83}$$

and to the compatibility equations in the form

$$\frac{d\varepsilon_r}{dp} = \xi_r, \quad \frac{d\varepsilon_\theta}{dp} = \xi_\theta, \quad \frac{d\varepsilon_z}{dp} = \xi_z. \tag{1.84}$$

It is understood here that ψ involved in ξ_r, ξ_θ and ξ_z in Eq. (1.84) is eliminated by means of the solution to Eq. (1.83).

1.4.4 Yield Criterion (1.11)

Substituting Eq. (1.53) into Eq. (1.20) yields

$$\begin{aligned}
\frac{\xi_r^e}{k} &= -\frac{\beta_1}{2}\left\{\begin{array}{c}\left[1+\nu+3\sqrt{3}(1-\nu)\beta_1\right]\cos\psi+\\ +\sqrt{3}\left[1+\nu-\sqrt{3}(1-\nu)\beta_1\right]\sin\psi\end{array}\right\}\frac{\partial\psi}{\partial p},\\
\frac{\xi_\theta^e}{k} &= \frac{\beta_1}{2}\left\{\begin{array}{c}\left[1+\nu-3\sqrt{3}(1-\nu)\beta_1\right]\cos\psi+\\ +\sqrt{3}\left[1+\nu+\sqrt{3}(1-\nu)\beta_1\right]\sin\psi\end{array}\right\}\frac{\partial\psi}{\partial p},\\
\frac{\xi_z^e}{k} &= 3\nu\beta_1^2\left(\sqrt{3}\cos\psi-\sin\psi\right)\frac{\partial\psi}{\partial p}
\end{aligned} \tag{1.85}$$

in the plastic zone. In the case of plastically incompressible material it is necessary to use the flow rule in the form of Eq. (1.17). Substituting Eq. (1.53) into Eq. (1.17) results in

$$\begin{aligned}
\xi_r^p &= \frac{3}{2}\lambda\sigma_0\left[\beta_1\left(\sqrt{3}-\beta_1\right)\cos\psi-\beta_1\left(1+\sqrt{3}\beta_1\right)\sin\psi+2\beta_0\right],\\
\xi_\theta^p &= \frac{3}{2}\lambda\sigma_0\left[\beta_1\left(1-\sqrt{3}\beta_1\right)\sin\psi-\beta_1\left(\sqrt{3}+\beta_1\right)\cos\psi+2\beta_0\right],\\
\xi_z^p &= 3\lambda\sigma_0\left[\beta_1^2\left(\cos\psi+\sqrt{3}\sin\psi\right)-2\beta_0\right].
\end{aligned}$$

Eliminating λ between these equations leads to

$$\begin{aligned}
\xi_r^p &= \xi_\theta^p\left[\frac{\beta_1\left(\sqrt{3}-\beta_1\right)\cos\psi-\beta_1\left(1+\sqrt{3}\beta_1\right)\sin\psi+2\beta_0}{\beta_1\left(1-\sqrt{3}\beta_1\right)\sin\psi-\beta_1\left(\sqrt{3}+\beta_1\right)\cos\psi+2\beta_0}\right],\\
\xi_z^p &= 2\xi_\theta^p\left[\frac{\beta_1^2\left(\cos\psi+\sqrt{3}\sin\psi\right)-2\beta_0}{\beta_1\left(1-\sqrt{3}\beta_1\right)\sin\psi-\beta_1\left(\sqrt{3}+\beta_1\right)\cos\psi+2\beta_0}\right].
\end{aligned} \tag{1.86}$$

In order to find the derivative $\partial\psi/\partial p$ involved in Eq. (1.85), it is necessary to differentiate Eq. (1.56). Then,

$$\frac{\left[\left(1+3\sqrt{3}\beta_1\right)\cos\psi+\sqrt{3}\left(1-\sqrt{3}\beta_1\right)\sin\psi\right]d\psi}{\left(\sqrt{3}\cos\psi-\sin\psi\right)}=$$
$$=\frac{\left[\left(1+3\sqrt{3}\beta_1\right)\cos\psi_0+\sqrt{3}\left(1-\sqrt{3}\beta_1\right)\sin\psi_0\right]d\psi_0}{\left(\sqrt{3}\cos\psi_0-\sin\psi_0\right)}+2\frac{d\rho}{\rho}.$$

It follows from this equation that

$$\frac{\partial\psi}{\partial p}=\frac{\Omega_P\left(\psi_0\right)}{\Omega_P\left(\psi\right)}\frac{d\psi_0}{dp},\tag{1.87}$$
$$\Omega_P\left(x\right)=\frac{\left(1+3\sqrt{3}\beta_1\right)\cos x+\sqrt{3}\left(1-\sqrt{3}\beta_1\right)\sin x}{4\left(\sqrt{3}\cos x-\sin x\right)}.$$

It is seen from Eqs. (1.54), (1.69) and (1.87) that $\Omega_P(x)\equiv\Omega_M(x)$ if $\alpha=0$. Eliminating ξ_r^p in (1.59) by means of Eq. (1.86) and, then, ξ_θ^p by means of Eqs. (1.22) and (1.58) lead to

$$\rho\frac{\partial\xi_\theta}{\partial\rho}+\frac{4\beta_1\sin\left(\psi-\pi/3\right)}{\left[2\beta_0-\beta_1\left(\sqrt{3}+\beta_1\right)\cos\psi+\left(1-\sqrt{3}\beta_1\right)\sin\psi\right]}\xi_\theta=$$
$$=\xi_r^e-\xi_\theta^e+\frac{4\beta_1\sin\left(\psi-\pi/3\right)}{\left[2\beta_0-\beta_1\left(\sqrt{3}+\beta_1\right)\cos\psi+\left(1-\sqrt{3}\beta_1\right)\sin\psi\right]}\xi_\theta^e+$$
$$+\frac{4k\beta_1\sin\left(\psi-\pi/3\right)}{\left[2\beta_0-\beta_1\left(\sqrt{3}+\beta_1\right)\cos\psi+\left(1-\sqrt{3}\beta_1\right)\sin\psi\right]}\frac{d\tau}{dp}.$$

Replacing here differentiation with respect to ρ with differentiation with respect to ψ by means of Eq. (1.55) and eliminating ξ_r^e and ξ_θ^e by means of Eqs. (1.85) and (1.87) result in

$$\frac{\partial\xi_\theta}{\partial\psi}=\xi_\theta W_0\left(\psi\right)+k\Omega_P\left(\psi_0\right)\frac{d\psi_0}{dp}W_1\left(\psi\right)-k\frac{d\tau}{dp}W_0\left(\psi\right)\tag{1.88}$$

where

$$W_0\left(\psi\right)=-\frac{\beta_1\left[\left(1+3\sqrt{3}\beta_1\right)\cos\psi+\sqrt{3}\left(1-\sqrt{3}\beta_1\right)\sin\psi\right]}{\left[\beta_1\left(\sqrt{3}+\beta_1\right)\cos\psi+\beta_1\left(\sqrt{3}\beta_1-1\right)\sin\psi-2\beta_0\right]},\tag{1.89}$$
$$W_1\left(\psi\right)=\frac{8\beta_1\left[\beta_1^2\left(\nu-2\right)+\beta_0\left(1+\nu\right)\sin\left(\psi+\frac{\pi}{6}\right)+\beta_1^2\left(2\nu-1\right)\cos\left(2\psi+\frac{\pi}{3}\right)\right]}{\left[\beta_1\left(\sqrt{3}+\beta_1\right)\cos\psi+\beta_1\left(\sqrt{3}\beta_1-1\right)\sin\psi-2\beta_0\right]}.$$

Equation (1.88) is a linear ordinary differential equation for ξ_θ. The solution to this equation satisfying the boundary condition (1.62) is

$$\frac{\xi_\theta}{k} = \left\{ \begin{array}{l} \Omega_P(\psi_0)\frac{d\psi_0}{dp}\int\limits_{\psi_c}^{\psi}\exp\left[-\int\limits_{\psi_c}^{\mu_1}W_0(\mu)\,d\mu\right]W_1(\mu_1)\,d\mu_1 - \\ -\frac{d\tau}{dp}\int\limits_{\psi_c}^{\psi}\exp\left[-\int\limits_{\psi_c}^{\mu_1}W_0(\mu)\,d\mu\right]W_0(\mu_1)\,d\mu_1 + \frac{\xi_c}{k} \end{array} \right\} \times \tag{1.90}$$

$$\times \exp\left[\int\limits_{\psi_c}^{\psi}W_0(\mu)\,d\mu\right].$$

Here both μ and μ_1 are dummy variables of integration. The value of ξ_c is given by Eq. (1.63). Using Eqs. (1.85), (1.87) and (1.90) it is possible to find ξ_θ^p from Eqs. (1.22) and (1.58) as $\xi_\theta^p = \xi_\theta - \xi_\theta^e - kd\tau/dp$. Then, ξ_r^p and ξ_z^p are determined from Eq. (1.86). Equations (1.22), (1.58), (1.85) and (1.87) supply ξ_r and ξ_z as functions of ψ and p. Equation (1.72) are valid. Eliminating the derivative $\partial\psi/\partial p$ by means of Eq. (1.87) leads to the characteristic equation in the form

$$\frac{d\psi}{d\psi_0} = \frac{\Omega_P(\psi_0)}{\Omega_P(\psi)} \tag{1.91}$$

and to the compatibility equations in the form

$$\frac{d\varepsilon_r}{dp} = \xi_r, \quad \frac{d\varepsilon_\theta}{dp} = \xi_\theta, \quad \frac{d\varepsilon_z}{dp} = \xi_z. \tag{1.92}$$

It is understood here that ψ involved in ξ_r, ξ_θ and ξ_z in Eq. (1.92) is eliminated by means of the solution to Eq. (1.91).

In the case of plastically compressible material Eqs. (1.19) and (1.53) combine to give

$$\xi_r^p = 9\lambda\sigma_0\left[3\beta_1\sin\left(\frac{\pi}{3}-\psi\right) - \sin\left(\frac{\pi}{6}+\psi\right)\right],$$
$$\xi_\theta^p = -9\lambda\sigma_0\left[3\beta_1\sin\left(\frac{\pi}{3}-\psi\right) + \sin\left(\frac{\pi}{6}+\psi\right)\right],$$
$$\xi_z^p = 6\lambda\sigma_0\beta_1^2\left[9\alpha + \left(9+2\alpha^2\right)\sin\left(\psi+\frac{\pi}{6}\right)\right].$$

Eliminating λ between these equations leads to

$$\xi_r^p = \xi_\theta^p\frac{[3\beta_1\sin(\psi-\pi/3)+\sin(\psi+\pi/6)]}{[\sin(\psi+\pi/6)-3\beta_1\sin(\psi-\pi/3)]}, \tag{1.93}$$
$$\xi_z^p = -\frac{2\xi_\theta^p\beta_1^2}{3}\frac{\left[9\alpha+\left(9+2\alpha^2\right)\sin(\psi+\pi/6)\right]}{[\sin(\psi+\pi/6)-3\beta_1\sin(\psi-\pi/3)]}.$$

Equations (1.85) and (1.87) are valid. Eliminating ξ_r^p in Eq. (1.59) by means of Eq. (1.93) and, then, ξ_θ^p by means of Eqs. (1.22) and (1.58) lead to

$$\rho\frac{\partial\xi_\theta}{\partial\rho}+\frac{6\beta_1\xi_\theta}{[3\beta_1+\tan(\psi+\pi/6)]}=\xi_r^e-\xi_\theta^e\frac{[3\beta_1\sin(\psi-\pi/3)+\sin(\psi+\pi/6)]}{[\sin(\psi+\pi/6)-3\beta_1\sin(\psi-\pi/3)]}+$$
$$+\frac{6k\beta_1}{[3\beta_1+\tan(\psi+\pi/6)]}\frac{d\tau}{dp}.$$

Replacing here differentiation with respect to ρ with differentiation with respect to ψ by means of Eq. (1.55) and eliminating ξ_r^e and ξ_θ^e by means of Eqs. (1.85) and (1.87) result in

$$\frac{\partial\xi_\theta}{\partial\psi}+3\beta_1\xi_\theta=k\Omega_P(\psi_0)\frac{d\psi_0}{dp}W_1(\psi)+3\beta_1k\frac{d\tau}{dp}\tag{1.94}$$

where

$$W_1(\psi)=\frac{\beta_1\left\{\begin{array}{c}\left[1+\nu-9\beta_1^2(1-\nu)\right]\cos 2\psi-\\-9\beta_1^2(1-\nu)\left(2-\sqrt{3}\sin 2\psi\right)-\\-(1+\nu)\left(2+\sqrt{3}\sin 2\psi\right)\end{array}\right\}}{3\beta_1\cos(\psi+\pi/6)+\sin(\psi+\pi/6)}.\tag{1.95}$$

Equation (1.94) is a linear ordinary differential equation for ξ_θ. The solution to Eq. (1.94) satisfying the boundary condition (1.62) is

$$\frac{\xi_\theta}{k}=\Omega_P(\psi_0)\frac{d\psi_0}{dp}\exp[3\beta_1(\psi_c-\psi)]\int_{\psi_c}^{\psi}\exp[3\beta_1(\mu-\psi_c)]W_1(\mu)\,d\mu+$$
$$+\frac{\xi_c}{k}\exp[3\beta_1(\psi_c-\psi)]+\frac{d\tau}{dp}\{1-\exp[3\beta_1(\psi_c-\psi)]\}.\tag{1.96}$$

Here μ is a dummy variable of integration. The value of ξ_c is given by Eq. (1.63). Using Eqs. (1.85), (1.87) and (1.96) it is possible to find ξ_θ^p from Eqs. (1.22) and (1.58) as $\xi_\theta^p=\xi_\theta-\xi_\theta^e-kd\tau/dp$. Then, ξ_r^p and ξ_z^p are determined from Eq. (1.93). Equations (1.22), (1.58), (1.85) and (1.87) supply ξ_r and ξ_z as functions of ψ and p. Equations (1.91) and (1.92) are valid. It is understood that ψ involved in ξ_r, ξ_θ and ξ_z in Eq. (1.92) is eliminated by means of the solution to Eq. (1.91).

References

1. Reid L (1997) Incorporating hole cold expansion to meet durability and damage tolerance airworthiness objectives. SAE International, SAE Paper No 972624
2. Ghorashi M, Daneshpazhooh M (2001) Limit analysis of variable thickness circular plates. Comput Struct 70:461–468
3. Durban D (1987) An exact solution for the internally pressurized, elastoplastic, strain-hardening, annular plate. Acta Mech 66:111–128
4. Guven U (1998) Elastic-plastic stress distribution in a rotating hyperbolic disk with rigid inclusion. Int J Mech Sci 40:97–109
5. Guven U (1998) Stress distribution in a linear hardening annular disk of variable thickness subjected to external pressure. Int J Mech Sci 40:589–601
6. Guven U, Altay O (2000) Elastic-plastic solid disk with nonuniform heat source subjected to external pressure. Int J Mech Sci 42:831–842
7. Eraslan AN (2002) Inelastic deformations of rotating variable thickness solid disks by Tresca and von Mises criteria. Int J Comput Eng Sci 3:89–101
8. Eraslan AN (2003) Elastoplastic deformations of rotating parabolic solid disks using Tresca's yield criterion. Eur J Mech A Solids 22:861–874
9. Eraslan AN (2003) Elastic-plastic deformations of rotating variable thickness annular disks with free, pressurized and radially constrained boundary conditions. Int J Mech Sci 45: 643–667
10. Eraslan AN, Orcan Y (2002) On the rotating elastic-plastic solid disks of variable thickness having concave profiles. Int J Mech Sci 44:1445–1466
11. Arslan E, Mack W, Eraslan AN (2008) Effect of a temperature cycle on a rotating elastic-plastic shaft. Acta Mech 195:129–140
12. You LH, Zhang JJ (1999) Elastic-plastic stresses in a rotating solid disk. Int J Mech Sci 41: 269–282
13. You LH, Tang YY, Zhang JJ, Zheng CY (2000) Numerical analysis of elastic-plastic rotating disks with arbitrary variable thickness and density. Int J Solids Struct 37:7809–7820
14. You XY, You LH, Zhang JJ (2004) A simple and efficient numerical method for determination of deformations and stresses in rotating solid shafts with non-linear strain-hardening. Commun Numer Methods Eng 20:689–697
15. Eraslan AN, Argeso A (2002) Limit angular velocities of variable thickness rotating disks. Int J Solids Struct 39:3109–3130
16. Debski R, Zyczkowski M (2002) On decohesive carrying capacity of variable-thickness annular perfectly plastic disks. Z Angew Math Mech 82:655–669
17. Vivio F, Vullo L (2010) Elastic-plastic analysis of rotating disks having non-linearly variable thickness: residual stresses by overspeeding and service stress state reduction. Ann Solid Struct Mech 1:87–102
18. Budiansky B, Mangasarian DL (1960) Plastic stress concentration at a circular hole in an infinite sheet subjected to equal biaxial tension. Trans ASME J Appl Mech 27:59–64
19. Papanastasiou P, Durban D (1997) Elastoplastic analysis of cylindrical cavity problems in geomaterials. Int J Numer Anal Mech Geomech 21:133–149
20. Durban D, Papanastasiou P (1997) Cylindrical cavity expansion and contraction in pressure sensitive geomaterials. Acta Mech 122:99–122
21. Bradford IDR, Durban D (1998) Stress and deformation fields around a cylindrical cavity embedded in a pressure-sensitive elastoplastic medium. Trans ASME J Appl Mech 65: 374–379
22. Chen PCT (1973) A comparison of flow and deformation theories in a radially stressed annular plate. Trans ASME J Appl Mech 40:283–287
23. Roberts SM, Hall FR, Bael AV, Hartley P, Pillinger I, Sturgess CEN, Houtte PV, Aernoudt E (1992) Benchmark tests for 3-D, elasto-plastic, finite-element codes for the modeling of metal forming processes. J Mater Process Technol 34:61–68

24. Helsing J, Jonsson A (2002) On the accuracy of benchmark tables and graphical results in the applied mechanics literature. Trans ASME J Appl Mech 69:88–90
25. Ball DL (1995) Elastic-plastic stress analysis of cold expanded fastener holes. Fat Fract Eng Mater Struct 18:47–63
26. Hill R (1950) The mathematical theory of plasticity. Clarendon Press, Oxford
27. Bland DR (1956) Elastoplastic thick-walled tubes of work-hardening material subject to internal and external pressures and to temperature gradients. J Mech Phys Solids 4:209–229
28. Rees DWA (1990) Autofrettage theory and fatigue life of open-ended cylinders. J Strain Anal Eng Des 25:109–121
29. Luxmoore AR, Light MF, Evans WT (1977) A comparison of finite-element and experimental studies on plane stress crack geometries. J Strain Anal Eng Des 12:208–216
30. Simo JC, Taylor RL (1986) A return mapping algorithm for plane stress elastoplasticity. Int J Numer Meth Eng 22:649–670
31. Jetteur P (1986) Implicit integration algorithm for elastoplasticity in plane stress analysis. Eng Comput 3:251–253
32. Kleiber M, Kowalczyk P (1996) Sensitivity analysis in plane stress elasto-plasticity and elasto-viscoplasticity. Comput Meth Appl Mech Eng 137:395–409
33. Valoroso N, Rosati L (2009) Consistent derivation of the constitutive algorithm for plane stress isotropic plasticity. Part 1: Theoretical formulation. Int J Solids Struct 46:74–91
34. Triantafyllou SP, Koumousis VK (2012) An hysteretic quadrilateral plane stress element. Arch Appl Mech 82:1675–1687
35. Rees DWA (2006) Basic engineering plasticity. Elsevier, Amsterdam
36. Yoshida S, Oguchi A, Nobuki M (1971) Influence of high hydrostatic pressure on the flow stress of copper polycrystals. Trans Jpn Inst Met 12:238–242
37. Spitzig WA, Sober RJ, Richmond O (1976) The effect of hydrostatic pressure on the deformation behavior of maraging and HY-80 steels and its implications for plasticity theory. Metall Trans 7A:1703–1710
38. Spitzig WA (1979) Effect of hydrostatic pressure on plastic-flow properties of iron single crystals. Acta Metall 27:523–534
39. Kao AS, Kuhn HA, Spitzig WA, Richmond O (1990) Influence of superimposed hydrostatic pressure on bending fracture and formability of a low carbon steel containing globular sulfides. Trans ASME J Eng Mater Technol 112(1):26–30
40. Wilson CD (2002) A critical reexamination of classical metal plasticity. Trans ASME J Appl Mech 69:63–68
41. Liu PS (2006) Mechanical behaviors of porous metals under biaxial tensile loads. Mater Sci Eng 422A:176–183
42. Drucker DC, Prager W (1952) Soil mechanics and plastic analysis for limit design. Q Appl Math 10:157–165
43. Alexandrov S, Alexandrova N (2001) Thermal effects on the development of plastic zones in thin axisymmetric plates. J Strain Anal Eng Des 36:169–176
44. Alexandrov S, Jeng Y-R, Lomakin E (2011) Effect of pressure-dependency of the yield criterion on the development of plastic zones and the distribution of residual stresses in thin annular disks. Trans ASME J Appl Mech 78:031012
45. Alexandrov SE, Lomakin EV, Jeng Y-R (2012) Solution of the thermoelasticplastic problem for a thin disk of plastically compressible material subject to thermal loading. Dokl Phys 57:136–139
46. Alexandrov S, Jeng Y-R, Lyamina E (2012) Influence of pressure-dependency of the yield criterion and temperature on residual stresses and strains in a thin disk. Struct Eng Mech 44:289–303
47. Pirumov A, Alexandrov S, Jeng Y-R (2013) Enlargement of a circular hole in a disc of plastically compressible material. Acta Mech 224:2965–2976
48. Alexandrov S, Jeng Y-R, Lomakin E (2014) An exact semi-analytic solution for residual stresses and strains within a thin hollow disc of pressure-sensitive material subject to thermal loading. Meccanica 49:775–794

49. Alexandrov S, Lyamina E, Jeng Y-R (2012) Design of an annular disc subject to thermomechanical loading. Math Prob Eng 2012, Article ID 709178
50. Wang Y-C, Alexandrov S, Jeng Y-R (2013) Effects of thickness variations on the thermal elastoplastic behavior of annular discs. Struct Eng Mech 47:839–856
51. Alexandrov S, Lyamina E, Jeng Y-R (2013) Plastic collapse of a thin annular disk subject to thermomechanical loading. Trans ASME J Appl Mech 80:051006
52. Alexandrov S, Wang Y-C, Aizikovich S (2014) Effect of temperature-dependent mechanical properties on plastic collapse of thin discs. Proc IMechE Part C: J Mech Eng Sci 228:2483–2487
53. Alexandrov S, Wang Y-C, Jeng Y-R (2014) Elastic-plastic stresses and strains in thin discs with temperature-dependent properties subject to thermal loading. J Therm Stresses 37:488–505
54. Alexandrov S, Pham C (2014) Plastic collapse mechanisms in thin disks subject to thermomechanical loading. Asia Pacific J Comput Eng 1:7
55. Alexandrov S, Mustafa Y (2014) A qualitative comparison of flow rules of pressure-dependent plasticity under plane stress conditions. J Eng Math 89:177–191
56. Alexandrova N, Alexandrov S (2004) Elastic-plastic stress distribution in a plastically anisotropic rotating disk. Trans ASME J Appl Mech 71:427–429

Chapter 2
Mechanical Loading

2.1 Disc Under Internal Pressure

The disc shown in Fig. 1.1 is loaded by internal pressure q_0 and its outer surface is stress free. Therefore, $\tau = 0$, $d\tau/dp = 0$, $\varepsilon_r^T = \varepsilon_\theta^T = \varepsilon_z^T = 0$, and $\xi_r^T = \xi_\theta^T = \xi_z^T = 0$. The boundary conditions are

$$\sigma_r = 0 \quad \text{for} \quad \rho = 1 \tag{2.1}$$

and

$$\sigma_r = -q_0 \quad \text{for} \quad \rho = a. \tag{2.2}$$

At the stage of purely elastic loading these boundary conditions and Eq. (1.29) combine to give

$$A + B = 0, \quad \frac{A}{a^2} + B = -q \tag{2.3}$$

where $q = q_0/\sigma_0$. Solving Eq. (2.3) for A and B yields

$$A = -B = -\frac{qa^2}{1 - a^2}. \tag{2.4}$$

This equation and one of the yield criteria combine to determine the value of q corresponding to the initiation of plastic yielding. This value of q will be denoted by q_e. The corresponding value of the function ψ involved in Eqs. (1.32), (1.47) and (1.53) will be denoted by ψ_e. It is understood here that ψ is calculated at the site of plastic yielding initiation and the plastic zone reduces to a circle at this instant. The solutions considered in this chapter are for elastic/plastic discs. Therefore, $q \geq q_e$ and there is an elastic plastic boundary, $\rho = \rho_c$, where $\psi = \psi_c$.

S. Alexandrov, *Elastic/Plastic Discs Under Plane Stress Conditions*,
SpringerBriefs in Computational Mechanics, DOI 10.1007/978-3-319-14580-8_2

It is worthy of note that the enlargement of a hole in plates or discs is one of the classical problems in plasticity. Solutions to this problem for various material models are contained in textbooks and monographs [1–4]. A recent review of available solutions for the enlargement of a circular hole in thin plates has been given in [5].

2.1.1 Yield Criterion (1.5)

Substituting Eq. (2.4) into Eq. (1.31) results in

$$q_e = \frac{1-a^2}{\sqrt{3+a^4}}. \tag{2.5}$$

Since the plastic zone starts to develop from the inner radius of the disc, the material is elastic in the range $\rho_c \le \rho \le 1$. Therefore, Eq. (1.29) are valid in this range. However, A and B are not determined by Eq. (2.4). Nevertheless, the radial stress determined by Eq. (1.29) must satisfy the boundary condition (2.1). Then,

$$A + B = 0. \tag{2.6}$$

The radial stress determined by Eq. (1.32) must satisfy the boundary condition (2.2). Therefore,

$$\frac{2}{\sqrt{3}} \sin \psi_a = q \tag{2.7}$$

where ψ_a is the value of ψ at $\rho = a$. It follows from Eqs. (1.27), (1.29) and (1.32) that

$$\frac{2 \sin \psi_c}{\sqrt{3}} = A\left(1 - \frac{1}{\rho_c^2}\right), \quad \frac{\sin \psi_c}{\sqrt{3}} + \cos \psi_c = A\left(1 + \frac{1}{\rho_c^2}\right). \tag{2.8}$$

Here B has been eliminated by means of Eq. (2.6). It is convenient to put $\psi_0 = \psi_a = p$ and $\rho_0 = a$ in Eqs. (1.16) and (1.35). Then, Eq. (1.35) becomes

$$\rho = a \exp\left[\frac{\sqrt{3}}{2}(\psi - \psi_a)\right]\sqrt{\frac{\sin(\psi_a - \pi/3)}{\sin(\psi - \pi/3)}}. \tag{2.9}$$

It follows from this equation and the definition for ψ_c that

$$\rho_c = a \exp\left[\frac{\sqrt{3}}{2}(\psi_c - \psi_a)\right]\sqrt{\frac{\sin(\psi_a - \pi/3)}{\sin(\psi_c - \pi/3)}}. \tag{2.10}$$

Solving Eq. (2.8) for ρ_c and A gives

$$\rho_c^2 = -\frac{\sqrt{3}\sin(\psi_c + \pi/6)}{\sin(\psi_c - \pi/3)}, \quad A = \frac{1}{2}\left(\sqrt{3}\sin\psi_c + \cos\psi_c\right). \tag{2.11}$$

The plastic zone starts to develop at $q = q_e$, $\rho_c = a$ and $\psi_a = \psi_c = \psi_e$. Then, it follows from Eqs. (2.5), (2.7) and (2.11) that ψ_e is found from the following equations

$$\frac{2}{\sqrt{3}}\sin\psi_e = \frac{1-a^2}{\sqrt{3+a^4}}, \quad -\frac{\sqrt{3}\sin(\psi_e + \pi/6)}{\sin(\psi_e - \pi/3)} = a^2. \tag{2.12}$$

These equations are compatible if $\cos(\pi/6 + \psi_e) < 0$. In this case, the value of ψ_e is determined from Eq. (2.12)[1] as

$$\psi_e = \pi - \arcsin\left[\frac{\sqrt{3}}{2}\frac{(1-a^2)}{\sqrt{3+a^4}}\right]. \tag{2.13}$$

Thus the value of ψ_e is in the range $5\pi/6 \le \psi_e < \pi$ when a varies in the range $0 \le a < 1$. Therefore, $\cos\psi_e < 0$ and it follows from Eq. (2.7) that $d\psi_a/dq < 0$ at $\psi_a = \psi_e$. Differentiating Eq. (2.11) for ρ_c with respect to q yields

$$\frac{d\rho_c^2}{dq} = \sqrt{3}\sec^2\left(\frac{\pi}{6} + \psi_c\right)\frac{d\psi_c}{dq}.$$

It is reasonable to assume that $d\rho_c/dq > 0$. Therefore, $d\psi_c/dq > 0$. Finally, it is possible to conclude that the value of ψ varies in the range

$$\psi_a \le \psi \le \psi_c. \tag{2.14}$$

The entire disc becomes plastic when $\rho_c = 1$. The corresponding value of ψ_c is denoted by ψ_q. It follows from Eq. (2.11) for ρ_c that the equation for ψ_q is $\sin\psi_q = 0$. Therefore,

$$\psi_q = \pi. \tag{2.15}$$

Since $\sin(\psi_c + \pi/6) \le 0$ and $\sin(\psi_c - \pi/3) > 0$ in the range $\psi_e \le \psi_c \le \psi_q$, substituting Eq. (2.11) for ρ_c into Eq. (2.10) leads to the following equation that connects ψ_a and ψ_c

$$a = \sqrt{\frac{-\sqrt{3}\sin(\psi_c + \pi/6)}{\sin(\psi_a - \pi/3)}}\exp\left[\frac{\sqrt{3}}{2}(\psi_a - \psi_c)\right]. \tag{2.16}$$

It is seen from Eq. (2.7) that $dq/d\psi_a = 0$ at $\psi_a = \pi/2$. The corresponding value of q is $q_m = 2/\sqrt{3}$. The plane stress approximation is not valid in the range $q > q_m$ since an intensive local thickening occurs in the vicinity of $\rho = a$. Therefore, the present solution is restricted to the range $\psi_e \geq \psi_a \geq \pi/2$. Another restriction follows from Eq. (2.15) and is $\psi_e \leq \psi_c \leq \pi$. Putting $\psi_a = \pi/2$ and $\psi_c = \pi$ in Eq. (2.16) gives

$$a_{cr} = \sqrt{\sqrt{3}\exp\left(-\frac{\sqrt{3}\pi}{4}\right)}. \tag{2.17}$$

Therefore, the present solution is restricted by the inequality $\psi_e \geq \psi_a \geq \pi/2$ if $a \leq a_{cr}$ and by the inequality $\psi_e \leq \psi_c \leq \pi$ if $a \geq a_{cr}$. Equation (2.16) should be solved numerically to find the dependence of ψ_c on ψ_a. Then, the radius of the elastic/plastic boundary is immediately determined from Eq. (2.11) and the value of q follows from Eq. (2.7). The value of A is found from Eq. (2.11). Then, B is given by Eq. (2.6). The distribution of stresses is determined from Eq. (1.29) in the range $\rho_c \leq \rho \leq 1$ and from Eqs. (1.32) and (2.9) in the range $a \leq \rho \leq \rho_c$. The latter is in parametric form with ψ being the parameter varying in the range shown in Eq. (2.14).

Replacing ψ_0 with ψ_a in Eq. (1.74) and using Eq. (1.69) lead to

$$\frac{\cos\psi d\psi}{\sqrt{3}\cos\psi - \sin\psi} = \frac{\cos\psi_a d\psi_a}{\sqrt{3}\cos\psi_a - \sin\psi_a}.$$

Integrating this equation gives

$$\frac{\sin(\psi_a - \pi/3)}{\sin(\psi - \pi/3)}\exp\left[\sqrt{3}(\psi - \psi_a)\right] = \frac{C^2}{a^2} \tag{2.18}$$

where C is constant on each characteristic curve. Comparing Eqs. (2.9) and (2.18) shows that $C = \rho$. Therefore, integrating Eq. (1.75) is equivalent to integrating at a fixed value of ρ. Two fields of characteristics found from Eq. (2.18) are shown in Fig. 2.1 for $a = 0.3 < a_{cr}$ and in Fig. 2.2 for $a = 0.5 > a_{cr}$. Since $p = \psi_a$, it is possible to rewrite Eq. (1.75) as

$$\frac{d\varepsilon_r}{d\psi_a} = \xi_r, \quad \frac{d\varepsilon_\theta}{d\psi_a} = \xi_\theta, \quad \frac{d\varepsilon_z}{d\psi_a} = \xi_z. \tag{2.19}$$

Differentiating Eq. (2.6) and Eq. (2.11) for A with respect to ψ_a gives

$$\frac{dA}{d\psi_a} = \frac{1}{2}\left(\sqrt{3}\cos\psi_c - \sin\psi_c\right)\frac{d\psi_c}{d\psi_a}, \quad \frac{dB}{d\psi_a} = -\frac{dA}{d\psi_a}. \tag{2.20}$$

The derivative $d\psi_c/d\psi_a$ is found from Eq. (2.16) as

$$\frac{d\psi_c}{d\psi_a} = \frac{a^2\cos(\psi_a - \pi/3)\exp\left[\sqrt{3}(\psi_c - \psi_a)\right] + 3\sin(\psi_c + \pi/6)}{2\sqrt{3}\sin\psi_c}. \tag{2.21}$$

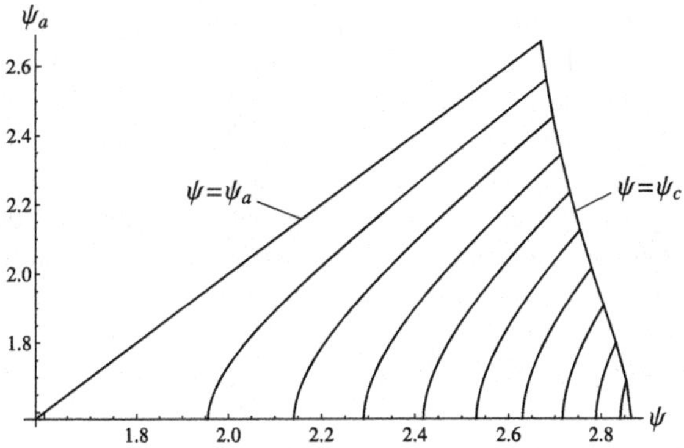

Fig. 2.1 Field of characteristics for an $a = 0.3$ disc

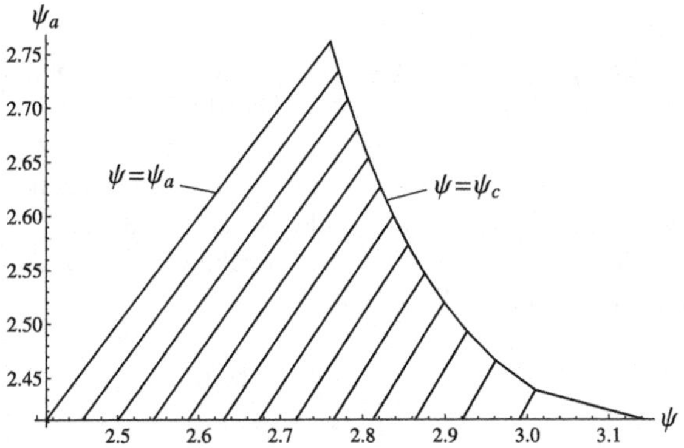

Fig. 2.2 Field of characteristics for an $a = 0.5$ disc

Using Eq. (2.11) for ρ_c and Eq. (2.20) it is possible to transform Eq. (1.63) to

$$\frac{\xi_c}{k} = \frac{\left(\sqrt{3}\cos\psi_c + \sin\psi_c - 2\nu\sin\psi_c\right)\sin\left(\psi_c - \pi/3\right)}{\sqrt{3}\sin\left(\psi_c + \pi/6\right)}\frac{d\psi_c}{d\psi_a}. \tag{2.22}$$

Equation (1.71) becomes

$$\frac{\xi_\theta}{k} = \frac{\xi_c}{k}\exp\left[\sqrt{3}\left(\psi_c - \psi\right)\right] + \frac{\cos\psi_a}{\sqrt{3}\left(\sqrt{3}\cos\psi_a - \sin\psi_a\right)}\times \tag{2.23}$$

$$\times\int\limits_{\psi_c}^{\psi}\frac{\left[(1-2\nu)\left(\sqrt{3}\sin 2\mu - \cos 2\mu\right) - 2\left(2-\nu\right)\right]}{\cos\mu}\exp\left[\sqrt{3}\left(\mu - \psi\right)\right]d\mu.$$

It has been taken into account here that $d\psi_0/dp = 1$ in the case under consideration. Using Eq. (2.21) the right hand side of Eq. (2.22) is represented as a function of ψ_a and ψ_c. Therefore, the right hand side of Eq. (2.23) depends on ψ, ψ_a and ψ_c. Eliminating ψ_c by means of the solution of Eq. (2.16) supplies the right hand side of Eq. (2.23) as a function of ψ and ψ_a. Further eliminating ψ by means of the solution to Eq. (2.18) at any given value of $C = \rho$ determines the right hand side of Eq. (2.19) for ε_θ as a function of ψ_a. This function is denoted by $E_\theta(\psi_a)$. Therefore, Eq. (2.19) for ε_θ can be integrated numerically. In particular, the value of ε_θ at $\psi_a = \psi_m$ and $\rho = C$ is given by

$$\varepsilon_\theta = \int_{\psi_i}^{\psi_m} E_\theta(\psi_a)\, d\psi_a + E_\theta^e. \tag{2.24}$$

Here ψ_m is prescribed and Eq. (2.24) supplies ε_θ in the plastic zone. It is seen from Eq. (2.7) that prescribing the value of ψ_m is equivalent to prescribing the value of q. The procedure to find ψ_i and E_θ^e is as follows. The value of E_θ^e is the circumferential strain at $\rho = \rho_c = C$ and the value of ψ_i is the value of ψ_a at $\rho_c = C$. It is seen from Eq. (1.26)[1] that E_θ^e is determined from the solution in the elastic zone. Alternatively, since the stresses are continuous across the elastic/plastic boundary, the value of E_θ^e can be found from Eqs. (1.1) and (1.32). Let ψ_C be the value of ψ_c at $\rho_c = C$. Then, the equation for ψ_C follows from Eq. (2.11) as

$$C^2 = -\frac{\sqrt{3}\sin(\psi_C + \pi/6)}{\sin(\psi_C - \pi/3)}. \tag{2.25}$$

Having found the value of ψ_C the value of E_θ^e is determined from Eqs. (1.1) to (1.32) at $\psi = \psi_C$ as

$$\frac{E_\theta^e}{k} = -\cos\psi_C - \frac{(1-2\nu)}{\sqrt{3}}\sin\psi_C. \tag{2.26}$$

Substituting $\psi_c = \psi_C$ into Eq. (2.16) and solving this equation for ψ_a supplies the value of ψ_i involved in Eq. (2.24).

The distributions of ε_r and ε_z in the plastic zone are determined in a similar manner. In particular, using Eqs. (1.22) and (1.68)

$$\begin{aligned} \xi_r &= \xi_r^e + \xi_r^p = \xi_r^e + \xi_\theta^p \frac{\sin(\psi - \pi/6)}{\cos\psi} = \xi_r^e + \left(\xi_\theta - \xi_\theta^e\right)\frac{\sin(\psi - \pi/6)}{\cos\psi}, \\ \xi_z &= \xi_z^e + \xi_z^p = \xi_z^e - \xi_\theta^p \frac{\sin(\psi + \pi/6)}{\cos\psi} = \xi_z^e - \left(\xi_\theta - \xi_\theta^e\right)\frac{\sin(\psi + \pi/6)}{\cos\psi}. \end{aligned} \tag{2.27}$$

Using Eqs. (1.66), (1.69) and (2.23) the right hand sides of Eq. (2.27) are expressed in terms of ψ and ψ_a. Eliminating ψ by means of the solution to Eq. (2.18) at a given value of C determines the right hand sides of these equations as functions of ψ_a.

These functions are denoted by $E_r(\psi_a)$ and $E_z(\psi_a)$. Equation (2.19) for ε_r and ε_z can be integrated numerically. In particular,

$$\varepsilon_r = \int_{\psi_i}^{\psi_m} E_r(\psi_a)\, d\psi_a + E_r^e, \quad \varepsilon_z = \int_{\psi_i}^{\psi_m} E_z(\psi_a)\, d\psi_a + E_z^e. \tag{2.28}$$

These equations supply ε_r and ε_z in the plastic zone. In Eq. (2.28), E_r^e and E_z^e are the radial and axial strains, respectively, at $\rho = \rho_c = C$. These strains are determined from Eq. (1.61) at $\rho = \rho_c = C$, (2.6) and (2.11) or from Eqs. (1.1) and (1.32) at $\psi = \psi_C$. As a result,

$$\frac{E_r^e}{k} = \nu \cos\psi_C - \frac{(2-\nu)}{\sqrt{3}} \sin\psi_C, \quad \frac{E_z^e}{k} = \nu\left(\sqrt{3}\sin\psi_C + \cos\psi_C\right). \tag{2.29}$$

Having found the distributions of the total strains in the plastic zone it is possible to determine their plastic portion by means of Eq. (1.3) in which the elastic strains should be eliminated using Eqs. (1.1), (1.32) and (2.9). The total strains in the elastic zone follow from Eq. (1.61) in which A and B should be eliminated by means of Eqs. (2.6) and (2.11). The value of ψ_c involved in Eq. (2.11) is determined from Eq. (2.16) assuming that $\psi_a = \psi_m$.

The solution is illustrated in Figs. 2.3, 2.4, 2.5 2.6 and 2.7 for an $a = 0.3 < a_{cr}$ disc and in Figs. 2.8, 2.9, 2.10, 2.11 and 2.12 for an $a = 0.5 > a_{cr}$ disc. In all calculations $\nu = 0.3$. The distributions of the radial and circumferential stresses are depicted in Figs. 2.3, 2.4, 2.8 and 2.9 for several values of ρ_c. The distributions of the radial, circumferential and axial strains are shown in Figs. 2.5, 2.6, 2.7, 2.10, 2.11 and 2.12 for the same values of ρ_c. The solid lines correspond to the total strains and the broken lines to the plastic strains.

2.1.2 Yield Criterion (1.8)

Substituting Eq. (2.4) into Eq. (1.38) yields

$$b_1 = \frac{q^2 a^4}{\left(1-a^2\right)^2}\left[1 + \eta_1(\eta_1 - \eta)\right] - \eta_1^2, \quad b_2 = \frac{2q^2 a^4}{\left(1-a^2\right)^2}\left(1 - \eta_1^2\right), \tag{2.30}$$

$$b_3 = \frac{q^2 a^4}{\left(1-a^2\right)^2}\left[1 + \eta_1(\eta_1 + \eta)\right].$$

It is convenient to consider the cases $\eta_1 \leq 1$ and $\eta_1 > 1$ separately. Firstly, it is assumed that $\eta_1 \leq 1$. In this case $b_2 \geq 0$ and the plastic zone starts to develop from the inner radius of the disc (see Sect. 1.3.3). Then, Eq. (1.37) at $\rho = a$ and Eq. (2.30) combine to give

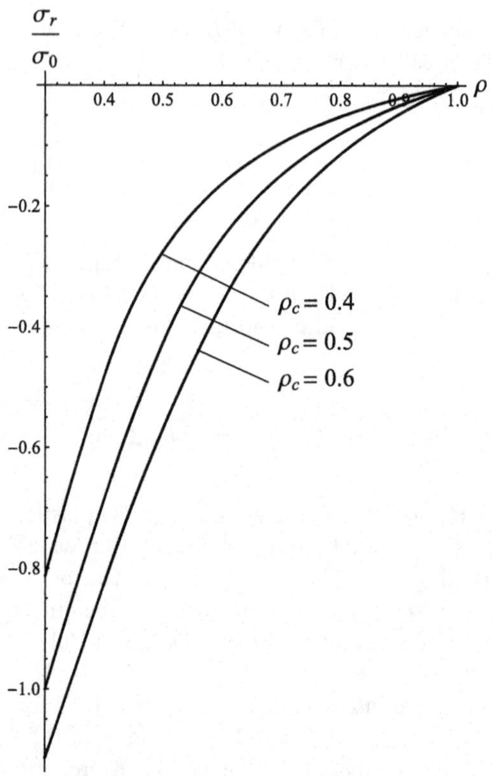

Fig. 2.3 Variation of the radial stress in an $a = 0.3$ disc at several values of ρ_c

$$q_e = \frac{\eta_1 \left(1 - a^2\right)}{\sqrt{[1 + \eta_1 (\eta_1 - \eta)]\, a^4 + 2\left(1 - \eta_1^2\right) a^2 + \eta_1 (\eta_1 + \eta) + 1}}. \tag{2.31}$$

Since the plastic zone starts to develop from the inner radius of the disc, the material is elastic in the range $\rho_c \le \rho \le 1$. Therefore, Eq. (1.29) are valid in this range. However, A and B are not determined by Eq. (2.4). Nevertheless, the radial stress from Eq. (1.29) must satisfy the boundary condition (2.1). Then,

$$A + B = 0. \tag{2.32}$$

The radial stress from Eq. (1.47) must satisfy the boundary condition (2.2). Therefore,

$$\frac{2}{\sqrt{4 - \eta^2}} \sin \psi_a = q \tag{2.33}$$

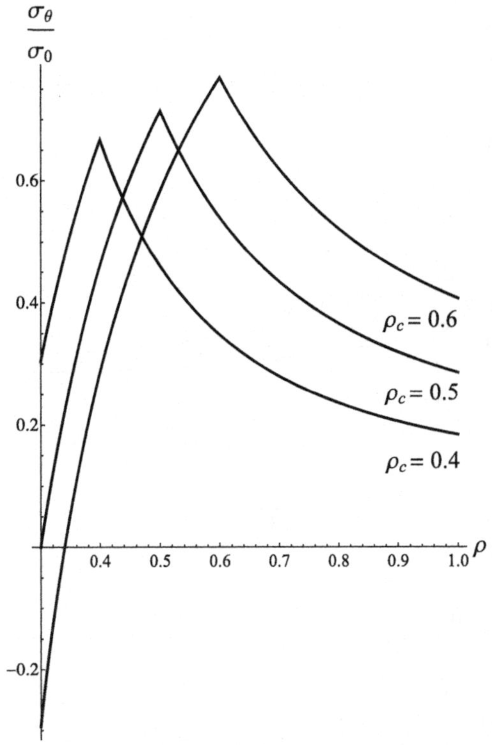

Fig. 2.4 Variation of the circumferential stress in an $a = 0.3$ disc at several values of ρ_c

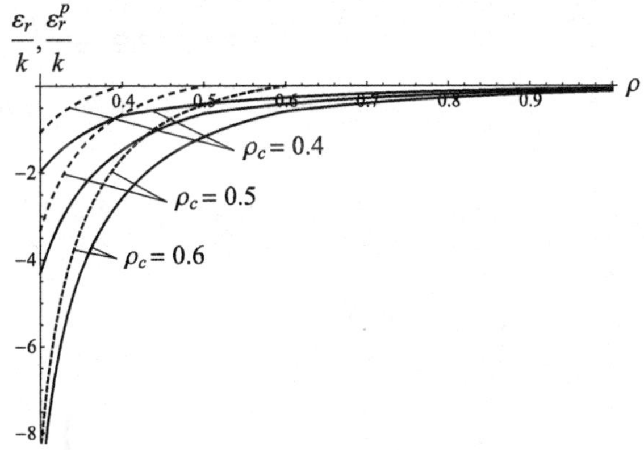

Fig. 2.5 Variation of the total and plastic radial strains in an $a = 0.3$ disc at several values of ρ_c

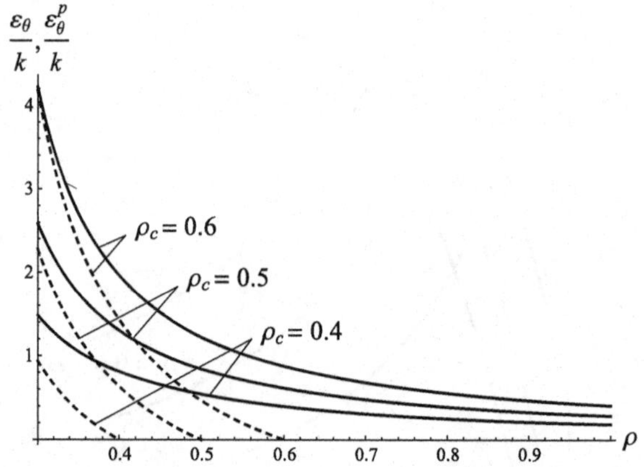

Fig. 2.6 Variation of the total and plastic circumferential strains in an $a = 0.3$ disc at several values of ρ_c

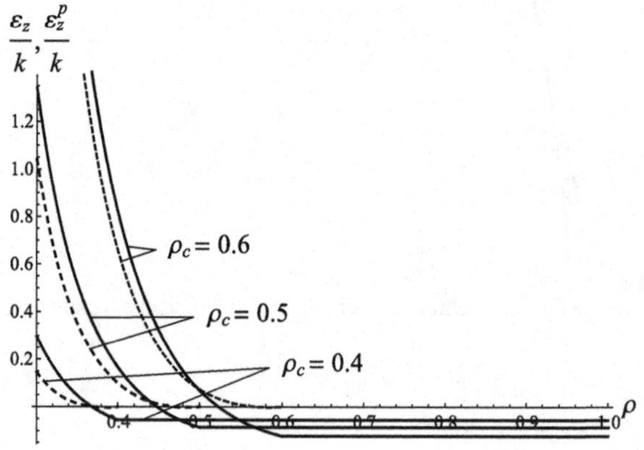

Fig. 2.7 Variation of the total and plastic axial strains in an $a = 0.3$ disc at several values of ρ_c

where ψ_a is the value of ψ at $\rho = a$. It follows from Eqs. (1.9), (1.27), (1.29), and (1.47) that

$$\left(1 - \frac{1}{\rho_c^2}\right) A = \frac{2 \sin \psi_c}{\sqrt{4 - \eta^2}}, \tag{2.34}$$

$$\left(\frac{1}{\rho_c^2} + 1\right) A = \eta_1 \left(\frac{\eta}{\sqrt{4 - \eta^2}} \sin \psi_c + \cos \psi_c\right).$$

It is convenient to put $\psi_0 = \psi_a = p$ and $\rho_0 = a$ in Eqs. (1.16) and (1.50). Then, Eq. (1.50) becomes

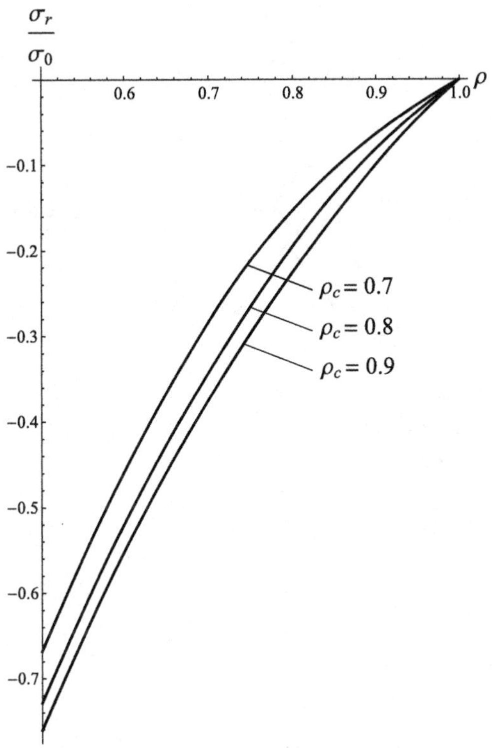

Fig. 2.8 Variation of the radial stress in an $a = 0.5$ disc at several values of ρ_c

$$\ln\left(\frac{\rho}{a}\right) = \frac{\eta_1\sqrt{4-\eta^2}}{2\left[1+\eta_1\left(\eta_1-\eta\right)\right]}\left(\psi-\psi_a\right) + \\ + \frac{\left(2-\eta\eta_1\right)}{2\left[1+\eta_1\left(\eta_1-\eta\right)\right]}\ln\left[\frac{\eta_1\sqrt{4-\eta^2}\cos\psi_a-\left(2-\eta\eta_1\right)\sin\psi_a}{\eta_1\sqrt{4-\eta^2}\cos\psi-\left(2-\eta\eta_1\right)\sin\psi}\right]. \tag{2.35}$$

It follows from this equation and the definition for ψ_c that

$$\ln\left(\frac{\rho_c}{a}\right) = \frac{\eta_1\sqrt{4-\eta^2}}{2\left[1+\eta_1\left(\eta_1-\eta\right)\right]}\left(\psi_c-\psi_a\right) + \\ + \frac{\left(2-\eta\eta_1\right)}{2\left[1+\eta_1\left(\eta_1-\eta\right)\right]}\ln\left[\frac{\eta_1\sqrt{4-\eta^2}\cos\psi_a-\left(2-\eta\eta_1\right)\sin\psi_a}{\eta_1\sqrt{4-\eta^2}\cos\psi_c-\left(2-\eta\eta_1\right)\sin\psi_c}\right]. \tag{2.36}$$

Solving Eq. (2.34) for ρ_c and A gives

$$\rho_c^2 = \frac{\eta_1\left(\eta+\sqrt{4-\eta^2}\cot\psi_c\right)+2}{\eta_1\left(\eta+\sqrt{4-\eta^2}\cot\psi_c\right)-2}, \quad A = \frac{\eta_1}{2}\cos\psi_c + \frac{\left(2+\eta\eta_1\right)}{2\sqrt{4-\eta^2}}\sin\psi_c. \tag{2.37}$$

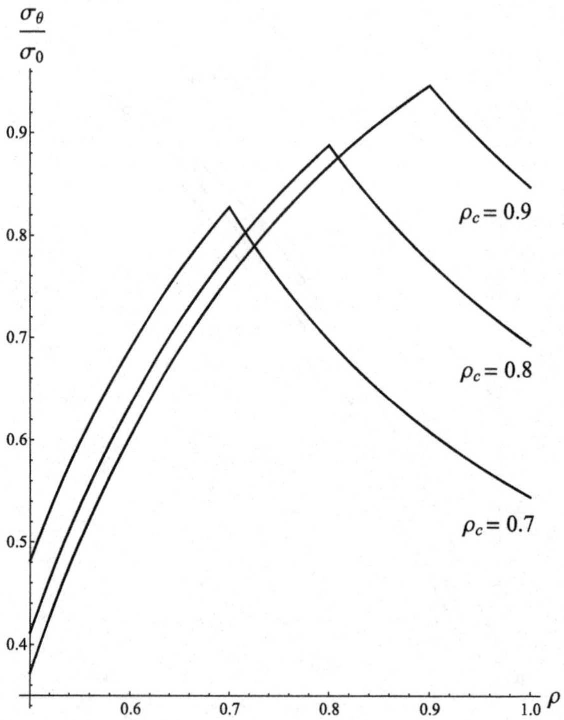

Fig. 2.9 Variation of the circumferential stress in an $a = 0.5$ disc at several values of ρ_c

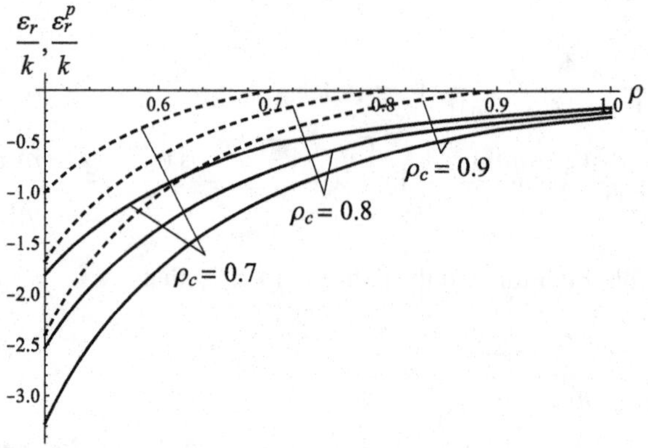

Fig. 2.10 Variation of the total and plastic radial strains in an $a = 0.5$ disc at several values of ρ_c

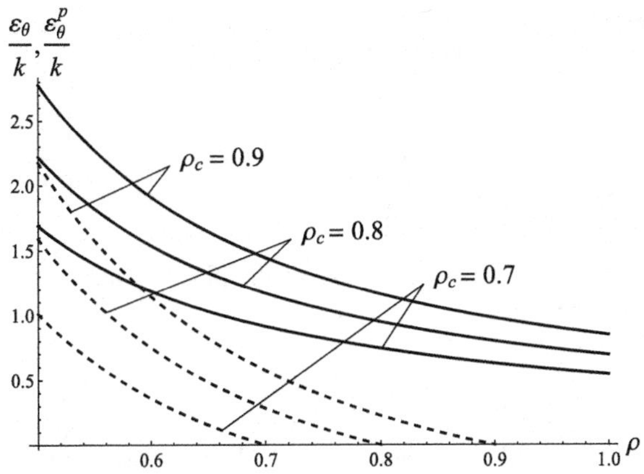

Fig. 2.11 Variation of the total and plastic circumferential strains in an $a = 0.5$ disc at several values of ρ_c

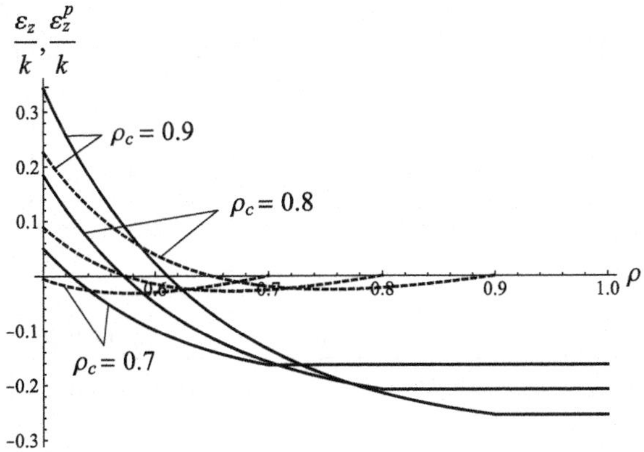

Fig. 2.12 Variation of the total and plastic axial strains in an $a = 0.5$ disc at several values of ρ_c

The plastic zone starts to develop at $q = q_e$, $\rho_c = a$ and $\psi_0 = \psi_a = \psi_e$. The value of ψ_e is determined from Eqs. (2.31) and (2.33) as

$$\psi_e = \pi - \arcsin\left[\frac{\eta_1\sqrt{4-\eta^2}\left(1-a^2\right)}{2\sqrt{\left[1+\eta_1\left(\eta_1-\eta\right)\right]a^4+2\left(1-\eta_1^2\right)a^2+\eta_1\left(\eta_1+\eta\right)+1}}\right]. \tag{2.38}$$

It has been taken into account here that $q > 0$. Thus the value of ψ_e is in the range $\psi_l \le \psi_e < \pi$ when a varies in the range $0 \le a < 1$. Here

$$\psi_l = \pi - \arcsin\left[\frac{\eta_1\sqrt{4-\eta^2}}{2\sqrt{\eta_1\left(\eta_1+\eta\right)+1}}\right]. \tag{2.39}$$

It is evident from Eq. (2.38) that $\cos\psi_e < 0$. Therefore, it follows from Eq. (2.33) that $d\psi_a/dq < 0$ at $\psi_a = \psi_e$. Differentiating Eq. (2.37) gives

$$\frac{d\rho_c^2}{dq} = \frac{4\eta_1\sqrt{4-\eta^2}}{\left(\eta\eta_1 - 2 + \eta_1\sqrt{4-\eta^2}\cot\psi_c\right)^2 \sin^2\psi_c}\frac{d\psi_c}{dq}. \tag{2.40}$$

Therefore, $d\psi_c/dq > 0$ as long as the size of the plastic zone increases. Thus it is reasonable to conclude that

$$\psi_a \le \psi \le \psi_c. \tag{2.41}$$

The entire disc becomes plastic when $\rho_c = 1$. The corresponding value of ψ_c is denoted by ψ_q. It follows from Eq. (2.34) that

$$\psi_q = \pi. \tag{2.42}$$

Eliminating ρ_c between Eqs. (2.36) and (2.37) supplies the equation that connects ψ_a and ψ_c in the form

$$\ln\left[\frac{\eta_1\left(\eta + \sqrt{4-\eta^2}\cot\psi_c\right) + 2}{\eta_1\left(\eta + \sqrt{4-\eta^2}\cot\psi_c\right) - 2}\right] - 2\ln a = \frac{\eta_1\sqrt{4-\eta^2}}{[1+\eta_1(\eta_1-\eta)]}(\psi_c - \psi_a) + \\
+ \frac{(2-\eta\eta_1)}{[1+\eta_1(\eta_1-\eta)]}\ln\left[\frac{\eta_1\sqrt{4-\eta^2}\cos\psi_a - (2-\eta\eta_1)\sin\psi_a}{\eta_1\sqrt{4-\eta^2}\cos\psi_c - (2-\eta\eta_1)\sin\psi_c}\right]. \tag{2.43}$$

It is seen from Eq. (2.33) that the maximum possible value of q is $q_m = 2/\sqrt{4-\eta^2}$ and that the present solution is restricted to the range $\psi_e \ge \psi_a \ge \pi/2$. Another restriction follows from Eq. (2.42) and is $\psi_e \le \psi_c \le \pi$. Putting $\psi_a = \pi/2$ and $\psi_c = \pi$ in Eq. (2.43) gives

$$\ln a_{cr} = -\frac{\pi\eta_1\sqrt{4-\eta^2}}{4[1+\eta_1(\eta_1-\eta)]} - \frac{(2-\eta\eta_1)}{2[1+\eta_1(\eta_1-\eta)]}\ln\left[\frac{2-\eta\eta_1}{\eta_1\sqrt{4-\eta^2}}\right]. \tag{2.44}$$

The present solution is restricted by the inequality $\psi_e \ge \psi_a \ge \pi/2$ if $a \le a_{cr}$ and by the inequality $\psi_e \le \psi_c \le \pi$ if $a \ge a_{cr}$. Equation (2.43) should be solved numerically to find the dependence of ψ_c on ψ_a. Then, the radius of the elastic/plastic boundary is immediately determined from Eq. (2.37)[1] and the value of q follows from Eq. (2.33). The values of A and B are found from Eqs. (2.32) and (2.37)[2]. The distribution of stresses is determined from Eq. (1.29) in the range $\rho_c \le \rho \le 1$ and from Eqs. (1.9), (1.47) to (2.35) in the range $a \le \rho \le \rho_c$. The latter is in parametric form with ψ being the parameter varying in the range shown in Eq. (2.41).

Replacing ψ_0 with ψ_a in Eq. (1.83) and using Eq. (1.79) lead to

$$\frac{d\psi}{\eta_1\sqrt{4-\eta^2}-(2-\eta\eta_1)\tan\psi}=\frac{d\psi_a}{\eta_1\sqrt{4-\eta^2}-(2-\eta\eta_1)\tan\psi_a}.$$

Integrating this equation gives

$$\begin{aligned}&\eta_1\sqrt{4-\eta^2}\,(\psi-\psi_a)+\\ +(2-\eta\eta_1)\ln&\left[\frac{\eta_1\sqrt{4-\eta^2}\cos\psi_a-(2-\eta\eta_1)\sin\psi_a}{\eta_1\sqrt{4-\eta^2}\cos\psi-(2-\eta\eta_1)\sin\psi}\right]=\\ &=2\left[1+\eta_1(\eta_1-\eta)\right]\ln\left(\frac{C}{a}\right)\end{aligned}\tag{2.45}$$

where C is constant on each characteristic curve. Comparing Eqs. (2.35) and (2.45) shows that $C=\rho$. Therefore, integrating along the characteristics is equivalent to integrating at fixed values of ρ. Since $p=\psi_a$, it is possible to rewrite Eq. (1.84) as

$$\frac{d\varepsilon_r}{d\psi_a}=\xi_r,\quad \frac{d\varepsilon_\theta}{d\psi_a}=\xi_\theta,\quad \frac{d\varepsilon_z}{d\psi_a}=\xi_z.\tag{2.46}$$

Differentiating Eqs. (2.32) and (2.37) for A with respect to ψ_a gives

$$\frac{dA}{d\psi_a}=\frac{1}{2}\left[\frac{(2+\eta\eta_1)}{\sqrt{4-\eta^2}}\cos\psi_c-\eta_1\sin\psi_c\right]\frac{d\psi_c}{d\psi_a},\quad \frac{dB}{d\psi_a}=-\frac{dA}{d\psi_a}.\tag{2.47}$$

The derivative $d\psi_c/d\psi_a$ is found from Eq. (2.43) as

$$\begin{aligned}\frac{d\psi_c}{d\psi_a}&=\frac{V_0(\psi_a)}{V_1(\psi_c)},\\ V_0(\psi_a)&=\frac{\eta_1\sqrt{4-\eta^2}}{1+\eta_1(\eta_1-\eta)}+\frac{(2-\eta\eta_1)}{[1+\eta_1(\eta_1-\eta)]}\times\\ &\times\frac{\eta_1\sqrt{4-\eta^2}\sin\psi_a+(2-\eta\eta_1)\cos\psi_a}{\eta_1\sqrt{4-\eta^2}\cos\psi_a-(2-\eta\eta_1)\sin\psi_a},\\ V_1(\psi_c)&=\frac{\eta_1\sqrt{4-\eta^2}}{1+\eta_1(\eta_1-\eta)}+\frac{(2-\eta\eta_1)}{[1+\eta_1(\eta_1-\eta)]}\times\\ &\times\frac{\eta_1\sqrt{4-\eta^2}\sin\psi_c+(2-\eta\eta_1)\cos\psi_c}{\eta_1\sqrt{4-\eta^2}\cos\psi_c-(2-\eta\eta_1)\sin\psi_c}-\\ &-\frac{4\eta_1\sqrt{4-\eta^2}}{\sin^2\psi_c\left[\eta_1^2\left(\eta+\sqrt{4-\eta^2}\cot\psi_c\right)^2-4\right]}.\end{aligned}\tag{2.48}$$

Using Eq. (2.37) for ρ_c and Eq. (2.47) it is possible to transform Eq. (1.63) to

$$\frac{\xi_c}{k} = \left[\eta_1\sqrt{4-\eta^2}\sin\psi_c - (2+\eta\eta_1)\cos\psi_c\right]\times \tag{2.49}$$
$$\times\frac{\left(\eta\eta_1 - 2\nu + \eta_1\sqrt{4-\eta^2}\cot\psi_c\right)}{\sqrt{4-\eta^2}\left(2+\eta\eta_1+\eta_1\sqrt{4-\eta^2}\cot\psi_c\right)}\frac{d\psi_c}{d\psi_a}$$

Substituting Eqs. (2.48) and (2.49) into Eq. (1.82) supplies the right hand side of this equation as a function of ψ, ψ_a and ψ_c. Then, ψ_c can be eliminated by means of the solution to Eq. (2.43). Further eliminating ψ by means of the solution to Eq. (2.45) at any given value of $C = \rho$ determines the right hand side of Eq. (2.46) for ε_θ as a function of ψ_a. This function is denoted by $E_\theta(\psi_a)$. Equation (2.46) for ε_θ can be integrated numerically. In particular, the value of ε_θ at $\psi_a = \psi_m$ and $\rho = C$ is given by

$$\varepsilon_\theta = \int_{\psi_i}^{\psi_m} E_\theta(\psi_a)\, d\psi_a + E_\theta^e. \tag{2.50}$$

Here ψ_m is prescribed and Eq. (2.50) supplies ε_θ in the plastic zone. It is seen from Eq. (2.33) that prescribing the value of ψ_m is equivalent to prescribing the value of q. The procedure to find ψ_i and E_θ^e is as follows. The value of E_θ^e is the circumferential strain at $\rho = \rho_c = C$ and the value of ψ_i is the value of ψ_a at $\rho_c = C$. It is seen from Eq. (1.26)[1] that E_θ^e is determined from the solution in the elastic zone. Alternatively, since the stresses are continuous across the elastic/plastic boundary, the value of E_θ^e can be found from Eqs. (1.1) and (1.47). Let ψ_C be the value of ψ_c at $\rho_c = C$. Then, the equation for ψ_C follows from Eq. (2.37) as

$$C^2 = \frac{\eta_1\left(\eta+\sqrt{4-\eta^2}\cot\psi_C\right)+2}{\eta_1\left(\eta+\sqrt{4-\eta^2}\cot\psi_C\right)-2}. \tag{2.51}$$

Having found the value of ψ_C the value of E_θ^e is determined from Eqs. (1.1) and (1.47) at $\psi = \psi_C$ as

$$\frac{E_\theta^e}{k} = -\frac{\eta_1\sqrt{4-\eta^2}\cos\psi_C + (\eta\eta_1 - 2\nu)\sin\psi_C}{\sqrt{4-\eta^2}}. \tag{2.52}$$

Substituting $\psi_c = \psi_C$ into Eq. (2.43) and solving this equation for ψ_a supplies the value of ψ_i involved in Eq. (2.50).

The distributions of ε_r and ε_z in the plastic zone are determined in a similar manner. In particular, using Eqs. (1.22) and (1.78)

$$\begin{aligned}\xi_r &= \xi_r^e + \xi_r^p = \xi_r^e + \frac{\xi_\theta^p \eta_1}{2}\left(\sqrt{4-\eta^2}\tan\psi - \eta\right) = \\ &= \xi_r^e + \frac{\eta_1}{2}\left(\xi_\theta - \xi_\theta^e\right)\left(\sqrt{4-\eta^2}\tan\psi - \eta\right), \\ \xi_z &= \xi_z^e + \xi_z^p = \xi_z^e - \frac{\xi_\theta^p}{2}\left(2 - \eta\eta_1 + \eta_1\sqrt{4-\eta^2}\tan\psi\right) = \\ &= \xi_z^e - \frac{\left(\xi_\theta - \xi_\theta^e\right)}{2}\left(2 - \eta\eta_1 + \eta_1\sqrt{4-\eta^2}\tan\psi\right).\end{aligned} \tag{2.53}$$

Using Eqs. (1.76), (1.79) and (1.82) the right hand sides of Eq. (2.53) are expressed in terms of ψ and ψ_a. Eliminating ψ by means of the solution to Eq. (2.45) at a given value of C determines the right hand sides of these equations as functions of ψ_a. These functions are denoted by $E_r(\psi_a)$ and $E_z(\psi_a)$. Equation (2.46) for ε_r and ε_z can be integrated numerically. In particular,

$$\varepsilon_r = \int_{\psi_i}^{\psi_m} E_r(\psi_a)\, d\psi_a + E_r^e, \quad \varepsilon_z = \int_{\psi_i}^{\psi_m} E_z(\psi_a)\, d\psi_a + E_z^e. \tag{2.54}$$

These equations supply ε_r and ε_z in the plastic zone. In Eq. (2.54), E_r^e and E_z^e are the radial and axial strains, respectively, at $\rho = \rho_c = C$. These strains are determined from Eq. (1.61) at $\rho = \rho_c = C$, (2.32) and (2.37) or from Eqs. (1.1) and (1.47) at $\psi = \psi_C$. As a result,

$$\begin{aligned}\frac{E_r^e}{k} &= \frac{\eta_1 \nu\sqrt{4-\eta^2}\cos\psi_C - (2+\nu\eta\eta_1)\sin\psi_C}{\sqrt{4-\eta^2}}, \\ \frac{E_z^e}{k} &= \nu\left[\eta_1\cos\psi_C + \frac{(2+\eta\eta_1)\sin\psi_C}{\sqrt{4-\eta^2}}\right].\end{aligned} \tag{2.55}$$

Having found the distributions of the total strains in the plastic zone it is possible to determine their plastic portion by means of Eq. (1.3) in which the elastic strains should be eliminated using Eqs. (1.1), (1.9), (1.47) and (2.35). The total strains in the elastic zone follow from Eq. (1.61) in which A and B should be eliminated by means of Eqs. (2.32) and (2.37). The value of ψ_c involved in Eq. (2.37) is determined from Eq. (2.43) assuming that $\psi_a = \psi_m$.

It is now assumed that $\eta_1 > 1$. The case corresponding to Eq. (1.44) is treated in the same manner as the case $\eta_1 \le 1$ since the plastic zone starts to develop from the inner radius of the disc. However, another plastic zone may start to develop from the outer radius of the disc if q is large enough. The corresponding condition follows from Eqs. (1.36), (2.32) and (2.37) in the form

$$(2-\eta\eta_1)\left[\eta_1\cos\psi_c+\frac{(2+\eta\eta_1)}{\sqrt{4-\eta^2}}\sin\psi_c\right]^2\le 2\eta_1^2. \tag{2.56}$$

If this inequality is not satisfied then it is necessary to find a solution with two plastic zones. This solution is beyond the scope of the present monograph.

Assume that Eq. (1.45) is satisfied. Then, the plastic zone starts to develop from the outer radius of the disc. Equation (1.37) at $\rho = 1$, (1.38) and (2.4) combine to give

$$q_e=\frac{\eta_1}{2}\frac{\left(1-a^2\right)}{a^2}. \tag{2.57}$$

The material is elastic in the range $a \le \rho \le \rho_c$. Therefore, Eq. (1.29) are valid in this range. However, A and B are not determined by Eq. (2.4). Nevertheless, the radial stress from Eq. (1.29) must satisfy the boundary condition (2.2). Then,

$$\frac{A}{a^2}+B=-q. \tag{2.58}$$

The radial stress determined by Eq. (1.47) must satisfy the boundary condition (2.1). Therefore,

$$\sin\psi=0 \tag{2.59}$$

at $\rho = 1$. It is evident that $\sigma_\theta > 0$ at $\rho = 1$. Then, it follows from Eqs. (1.9), (1.47) and (2.59) that

$$\psi=\pi \tag{2.60}$$

at $\rho = 1$. Therefore, it is convenient to put $\rho_0 = 1$ and $\psi_0 = \pi$ in Eq. (1.50). Then,

$$\ln\rho=\frac{\eta_1\sqrt{4-\eta^2}}{2\left[1+\eta_1\left(\eta_1-\eta\right)\right]}\left(\psi-\pi\right)+ \\ +\frac{(2-\eta\eta_1)}{2\left[1+\eta_1\left(\eta_1-\eta\right)\right]}\ln\left[\frac{\eta_1\sqrt{4-\eta^2}}{(2-\eta\eta_1)\sin\psi-\eta_1\sqrt{4-\eta^2}\cos\psi}\right]. \tag{2.61}$$

It follows from this equation that the radius of the elastic/plastic boundary is

$$\ln\rho_c=\frac{\eta_1\sqrt{4-\eta^2}}{2\left[1+\eta_1\left(\eta_1-\eta\right)\right]}\left(\psi_c-\pi\right)+ \\ +\frac{(2-\eta\eta_1)}{2\left[1+\eta_1\left(\eta_1-\eta\right)\right]}\ln\left[\frac{\eta_1\sqrt{4-\eta^2}}{(2-\eta\eta_1)\sin\psi_c-\eta_1\sqrt{4-\eta^2}\cos\psi_c}\right]. \tag{2.62}$$

It follows from Eqs. (1.27), (1.29) and (1.47) that

$$\left(\frac{1}{\rho_c^2} - \frac{1}{a^2}\right) A = -\frac{2 \sin \psi_c}{\sqrt{4-\eta^2}} + q,$$
$$\left(\frac{1}{\rho_c^2} + \frac{1}{a^2}\right) A = \eta_1 \left(\frac{\eta \sin \psi_c}{\sqrt{4-\eta^2}} + \cos \psi_c\right) - q.$$

Solving these equations for A and q gives

$$A = \frac{\rho_c^2}{2}\left[\frac{(\eta\eta_1 - 2)}{\sqrt{4-\eta^2}} \sin \psi_c + \eta_1 \cos \psi_c\right],$$
$$q = \frac{(\eta\eta_1 + 2) \sin \psi_c + \eta_1 \sqrt{4-\eta^2} \cos \psi_c}{2\sqrt{4-\eta^2}} + \tag{2.63}$$
$$+ \frac{\left[(2 - \eta\eta_1) \sin \psi_c - \eta_1 \sqrt{4-\eta^2} \cos \psi_c\right]}{2\sqrt{4-\eta^2}} \left(\frac{\rho_c}{a}\right)^2.$$

In these equations ρ_c can be eliminated by means of Eq. (2.62). Therefore, Eq. (2.63) supplies A and q as functions of ψ_c. The distribution of stresses is determined from Eq. (1.29) in the range $a \le \rho \le \rho_c$ and from Eqs. (1.9), (1.47) and (2.61) in the range $\rho_c \le \rho \le 1$. The latter is in parametric form with ψ being the parameter varying in the range $\psi_c \le \psi \le \pi$. The full disc becomes plastic when $\rho_c = a$. The corresponding values of ψ_c and q are determined from Eqs. (2.62) and (2.63).

It is convenient to put $p = \psi_c$ in Eq.(1.16). It is seen from Eq. (2.61) that ψ is independent of ψ_c. Therefore, it is evident from Eqs. (1.20) and (1.47) that $\xi_r^e = \xi_\theta^e = \xi_z^e = 0$ in the plastic zone. Thus $\xi_r^p = \xi_r$, $\xi_\theta^p = \xi_\theta$ and $\xi_z^p = \xi_z$. Moreover, $d\psi_0/dp = 0$ since $\psi_0 = \pi$. Therefore, using Eq. (1.81) it is possible to represent Eq. (1.80) as

$$\frac{\partial \xi_\theta}{\partial \psi} = \frac{\left[\sqrt{4-\eta^2}\eta_1 \tan \psi - \eta\eta_1 - 2\right]}{\left[\sqrt{4-\eta^2}\eta_1 + (\eta\eta_1 - 2) \tan \psi\right]} \xi_\theta. \tag{2.64}$$

Eliminating B in Eq. (1.63) by means of Eq. (2.58) results in

$$\frac{\xi_c}{k} = -\frac{dA}{d\psi_c}\left(\frac{1+\nu}{\rho_c^2} + \frac{1-\nu}{a^2}\right) - \frac{dq}{d\psi_c}(1-\nu). \tag{2.65}$$

Differentiating Eq. (2.63) with respect to ψ_c yields

$$2\sqrt{4-\eta^2}\frac{dA}{d\psi_c} = \left[(\eta\eta_1 - 2)\sin\psi_c + \eta_1\sqrt{4-\eta^2}\cos\psi_c\right]\frac{d\left(\rho_c^2\right)}{d\psi_c} + \\ + \rho_c^2\left[(\eta\eta_1 - 2)\cos\psi_c - \eta_1\sqrt{4-\eta^2}\sin\psi_c\right],$$

$$2\sqrt{4-\eta^2}\frac{dq}{d\psi_c} = (\eta\eta_1 + 2)\cos\psi_c - \eta_1\sqrt{4-\eta^2}\sin\psi_c + \\ + \left(\frac{\rho_c}{a}\right)^2\left[(2-\eta\eta_1)\cos\psi_c + \eta_1\sqrt{4-\eta^2}\sin\psi_c\right] + \\ + \frac{\left[(2-\eta\eta_1)\sin\psi_c - \eta_1\sqrt{4-\eta^2}\cos\psi_c\right]}{a^2}\frac{d\left(\rho_c^2\right)}{d\psi_c}. \tag{2.66}$$

Differentiating Eq. (2.62) with respect to ψ_c results in

$$\frac{d\left(\rho_c^2\right)}{\rho_c^2 d\psi_c} = \frac{\eta_1\sqrt{4-\eta^2}}{1+\eta_1(\eta_1-\eta)} - \\ - \frac{(2-\eta\eta_1)}{[1+\eta_1(\eta_1-\eta)]}\frac{\left[(2-\eta\eta_1)\cos\psi_c + \eta_1\sqrt{4-\eta^2}\sin\psi_c\right]}{\left[(2-\eta\eta_1)\sin\psi_c - \eta_1\sqrt{4-\eta^2}\cos\psi_c\right]}. \tag{2.67}$$

Using Eqs. (2.62), (2.66) and (2.67) the right hand side of Eq. (2.65) is found as a function of ψ_c. The solution of Eq. (2.64) satisfying the boundary condition (1.62) is

$$\frac{\xi_\theta}{k} = \frac{\xi_\theta^p}{k} = \frac{\xi_c(\psi_c)}{k}\exp\left[\frac{\sqrt{4-\eta^2}\eta_1(\psi_c-\psi)}{1+\eta_1(\eta_1-\eta)}\right] \times \\ \times \left[\frac{\eta_1\sqrt{4-\eta^2}\cos\psi_c + (\eta\eta_1-2)\sin\psi_c}{\eta_1\sqrt{4-\eta^2}\cos\psi + (\eta\eta_1-2)\sin\psi}\right]^{\eta_2} \tag{2.68}$$

where

$$\eta_2 = \frac{\eta_1^2 - 1}{1+\eta_1(\eta_1-\eta)}.$$

The procedure for finding the strains is as follows. Let Υ be the value of ρ at which the strains should be calculated at $q = q_m$. Eliminating ρ_c in Eq. (2.62) by means of Eq. (2.63) and putting $q = q_m$ leads to

$$\ln\left[\frac{(2+\eta\eta_1)\sin\psi_m+\eta_1\sqrt{4-\eta^2}\cos\psi_m-2q_m\sqrt{4-\eta^2}}{\eta_1\left(\eta\sin\psi_m+\sqrt{4-\eta^2}\cos\psi_m\right)-2\sin\psi_m}\right]+2\ln a=$$
$$=\frac{\eta_1\sqrt{4-\eta^2}\,(\psi_m-\pi)}{[1+\eta_1(\eta_1-\eta)]}+\frac{(2-\eta\eta_1)}{[1+\eta_1(\eta_1-\eta)]}\times \tag{2.69}$$
$$\times\ln\left[\frac{\eta_1\sqrt{4-\eta^2}}{(2-\eta\eta_1)\sin\psi_m-\eta_1\sqrt{4-\eta^2}\cos\psi_m}\right].$$

Here ψ_m is the value of ψ_c at $q=q_m$. Equation (2.69) should be solved for ψ_m numerically. Then, the value of ρ_c at $q=q_m$ denoted by ρ_m is determined from Eq. (2.63) as

$$\left(\frac{\rho_m}{a}\right)^2=\frac{(2+\eta\eta_1)\sin\psi_m+\eta_1\sqrt{4-\eta^2}\cos\psi_m-2q_m\sqrt{4-\eta^2}}{\eta_1\left(\eta\sin\psi_m+\sqrt{4-\eta^2}\cos\psi_m\right)-2\sin\psi_m}. \tag{2.70}$$

In the elastic region, $a\le\Upsilon\le\rho_m$, the distributions of the strains follow from Eq. (1.60) in the form

$$\frac{\varepsilon_r^e}{k}=\frac{\varepsilon_r}{k}=\frac{A_m(1+\nu)}{\Upsilon^2}+B_m(1-\nu),\quad \frac{\varepsilon_\theta^e}{k}=\frac{\varepsilon_\theta}{k}=-\frac{A_m(1+\nu)}{\Upsilon^2}+B_m(1-\nu),$$
$$\frac{\varepsilon_z^e}{k}=\frac{\varepsilon_z}{k}=-2\nu B_m. \tag{2.71}$$

Having found ψ_m and ρ_m from Eqs. (2.69) and (2.70) the values of A_m and B_m are determined by means of Eqs. (2.58) and (2.63) as

$$A_m=\frac{\rho_m^2}{2}\left[\frac{(\eta\eta_1-2)}{\sqrt{4-\eta^2}}\sin\psi_m+\eta_1\cos\psi_m\right], \tag{2.72}$$
$$B_m=-q_m-\frac{\rho_m^2}{2a^2}\left[\frac{(\eta\eta_1-2)}{\sqrt{4-\eta^2}}\sin\psi_m+\eta_1\cos\psi_m\right].$$

In order to calculate the strains in the plastic region, $\rho_m\le\Upsilon\le 1$, it is convenient to introduce the value of ψ_c at $\rho_c=\Upsilon$. This value of ψ_c is denoted by ψ_Υ. The equation for ψ_Υ follows from Eq. (2.62) in the form

$$2\ln\Upsilon=\frac{\eta_1\sqrt{4-\eta^2}\,(\psi_\Upsilon-\pi)}{[1+\eta_1(\eta_1-\eta)]}+ \tag{2.73}$$
$$+\frac{(2-\eta\eta_1)}{[1+\eta_1(\eta_1-\eta)]}\ln\left[\frac{\eta_1\sqrt{4-\eta^2}}{(2-\eta\eta_1)\sin\psi_\Upsilon-\eta_1\sqrt{4-\eta^2}\cos\psi_\Upsilon}\right].$$

This equation should be solved for ψ_Υ numerically. Then, the corresponding values of $q = q_\Upsilon$, $A = A_\Upsilon$ and $B = B_\Upsilon$ are found from Eqs. (2.58) and (2.63) as

$$2q_\Upsilon\sqrt{4-\eta^2} = (2+\eta\eta_1)\sin\psi_\Upsilon + \eta_1\sqrt{4-\eta^2}\cos\psi_\Upsilon - \left[\eta_1\left(\eta\sin\psi_\Upsilon + \sqrt{4-\eta^2}\cos\psi_\Upsilon\right) - 2\sin\psi_\Upsilon\right]\left(\frac{\Upsilon}{a}\right)^2, \tag{2.74}$$

$$A_\Upsilon = \frac{\Upsilon^2}{2}\left[\frac{(\eta\eta_1-2)}{\sqrt{4-\eta^2}}\sin\psi_\Upsilon + \eta_1\cos\psi_\Upsilon\right],$$

$$B_\Upsilon = -q_\Upsilon - \frac{\Upsilon^2}{2a^2}\left[\frac{(\eta\eta_1-2)}{\sqrt{4-\eta^2}}\sin\psi_\Upsilon + \eta_1\cos\psi_\Upsilon\right].$$

The elastic portions of the strains are determined from Eq. (1.60) in the form

$$\frac{\varepsilon_r^e}{k} = \frac{A_\Upsilon(1+\nu)}{\Upsilon^2} + B_\Upsilon(1-\nu), \quad \frac{\varepsilon_\theta^e}{k} = -\frac{A_\Upsilon(1+\nu)}{\Upsilon^2} + B_\Upsilon(1-\nu), \quad \frac{\varepsilon_z^e}{k} = -2\nu B_\Upsilon. \tag{2.75}$$

The plastic portions are given by

$$\varepsilon_r^p = \int_{\psi_\Upsilon}^{\psi_m}\xi_r^p\,d\psi_c, \quad \varepsilon_\theta^p = \int_{\psi_\Upsilon}^{\psi_m}\xi_\theta^p\,d\psi_c, \quad \varepsilon_z^p = \int_{\psi_\Upsilon}^{\psi_m}\xi_z^p\,d\psi_c. \tag{2.76}$$

Substituting Eq. (2.68) into Eq. (2.76) for ε_θ^p yields

$$\frac{\varepsilon_\theta^p}{k} = \int_{\psi_\Upsilon}^{\psi_m} \begin{matrix} \frac{\xi_c(\psi_c)}{k}\exp\left[\frac{\sqrt{4-\eta^2}\eta_1(\psi_c-\psi_\Upsilon)}{1+\eta_1(\eta_1-\eta)}\right]\times \\ \left[\frac{\eta_1\sqrt{4-\eta^2}\cos\psi_c + (\eta\eta_1-2)\sin\psi_c}{\eta_1\sqrt{4-\eta^2}\cos\psi_\Upsilon + (\eta\eta_1-2)\sin\psi_\Upsilon}\right]^{\eta_2} d\psi_c \end{matrix} \tag{2.77}$$

Eliminating ξ_r^p and ξ_z^p in Eq. (2.76) by means of Eq. (1.78) and taking into account that ψ is independent of ψ_c lead to

$$\varepsilon_r^p = \frac{\eta_1}{2}\left(\sqrt{4-\eta^2}\tan\psi_\Upsilon - \eta\right)\varepsilon_\theta^p, \tag{2.78}$$

$$\varepsilon_z^p = -\frac{\left(2-\eta\eta_1+\eta_1\sqrt{4-\eta^2}\tan\psi_\Upsilon\right)}{2}\varepsilon_\theta^p.$$

Here the strain ε_θ^p can be eliminated by means of Eq. (2.77). Using the solution to Eqs. (2.73) and (2.74) the total strains in the plastic zone are determined from Eqs. (1.3), (2.75), (2.77), and (2.78) at any given value of Υ.

A restriction on the solution found is that another plastic zone can start to develop from the inner radius of the disc. Substituting Eq. (2.58) into Eq. (1.36) at $\rho = a$ yields

$$[1 + \eta_1 (\eta_1 - \eta)] q^2 a^4 + 2A (2 - \eta\eta_1) q a^2 + 4A^2 = \eta_1^2 a^4. \tag{2.79}$$

Using Eqs. (2.62) and (2.63) the left hand side of Eq. (2.79) is expressed as a function of ψ_c. Therefore, this equation can be in general solved for ψ_c. If the solution to Eq. (2.79) exists and the corresponding value of ρ_c is in the range $a < \rho_c < 1$ then the domain of validity of the solution presented has been found. The corresponding value of q is denoted by q_m. A solution with two plastic zones is required in the range $q > q_m$. A similar solution is required from the beginning of plastic yielding if Eq. (1.46) is satisfied. Such solutions are beyond the scope of the present monograph.

Anisotropic coefficients involved in Eq. (1.6) are available in the literature. For example [6, 7],

$$\frac{F}{G+H} = 0.243, \quad \frac{H}{G+H} = 0.703 \quad \text{for steel DC06,}$$
$$\frac{F}{G+H} = 0.587, \quad \frac{H}{G+H} = 0.41 \quad \text{for aluminum alloy AA6016,}$$
$$\frac{F}{G+H} = 0.498, \quad \frac{H}{G+H} = 0.419 \quad \text{for aluminum alloy AA5182,}$$
$$\frac{F}{G+H} = 0.239, \quad \frac{H}{G+H} = 0.301 \quad \text{for aluminum alloy AA3104.}$$

Using these relations along with Eqs. (1.7) and (1.9) it is possible to determine the values of η and η_1. It has been checked by means of the procedure described in Sect. 1.3.3 that the plastic zone starts to develop from the inner radius of the disc in all these cases. The example given below is for the aluminum alloy AA3104. In this case, $\eta = 0.82$ and $\eta_1 = 0.735$. The fields of characteristics are shown in Fig. 2.13 for an $a = 0.3 < a_{cr}$ disc and in Fig. 2.14 for an $a = 0.5 > a_{cr}$ disc. The solution for stress and strain is illustrated in Figs. 2.15, 2.16, 2.17, 2.18 and 2.19 for an $a = 0.3$ disc and in Figs. 2.20, 2.21, 2.22, 2.23 and 2.24 for an $a = 0.5$ disc. The distributions of the radial and circumferential stresses are depicted in Figs. 2.15, 2.16, 2.20, and 2.21 for several values of ρ_c. The distributions of the radial, circumferential and axial strains are shown in Figs. 2.17, 2.18, 2.19, 2.22, 2.23 and 2.24 for the same values of ρ_c. The solid lines correspond to the total strains and the broken lines to the plastic strains. In all calculations $\nu = 0.3$.

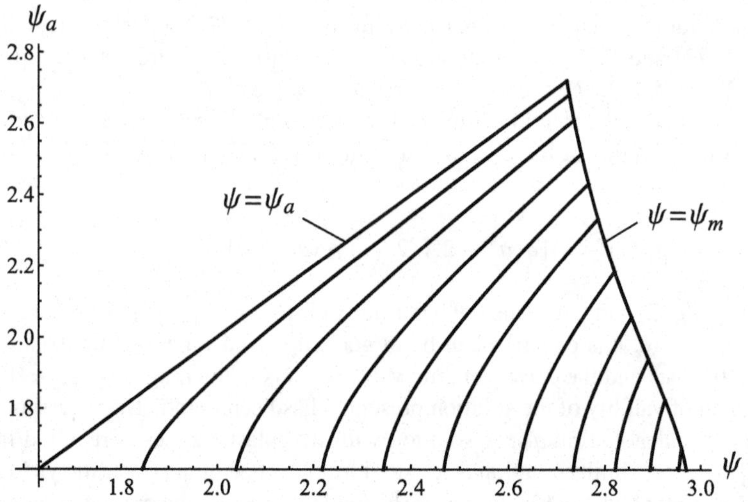

Fig. 2.13 Field of characteristics for an $a = 0.3$ disc at $\eta = 0.82$ and $\eta_1 = 0.735$

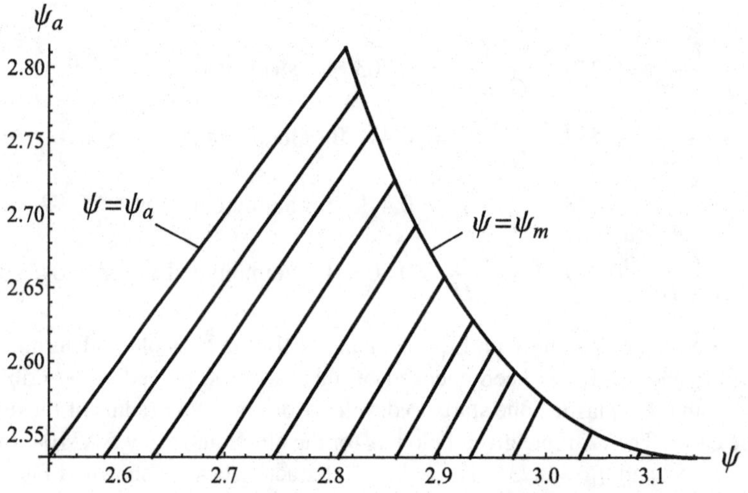

Fig. 2.14 Field of characteristics for an $a = 0.5$ disc at $\eta = 0.82$ and $\eta_1 = 0.735$

2.1.3 Yield Criterion (1.11)

Substituting Eq. (2.4) into Eq. (1.52) results in

$$q_e = \frac{3\left(1 - a^2\right)\left(3\sqrt{3 + a^4} - 2\alpha a^2\right)}{27 + a^4\left(9 - 4\alpha^2\right)}. \tag{2.80}$$

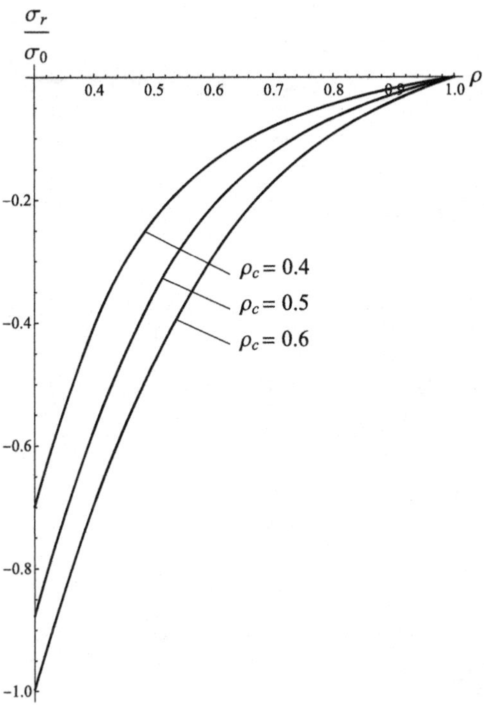

Fig. 2.15 Variation of the radial stress in an $a = 0.3$ disc at $\eta = 0.82$, $\eta_1 = 0.735$ and several values of ρ_c

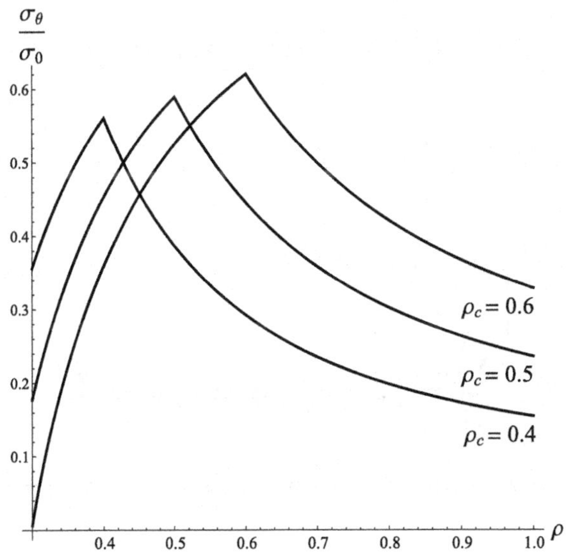

Fig. 2.16 Variation of the circumferential stress in an $a = 0.3$ disc at $\eta = 0.82$, $\eta_1 = 0.735$ and several values of ρ_c

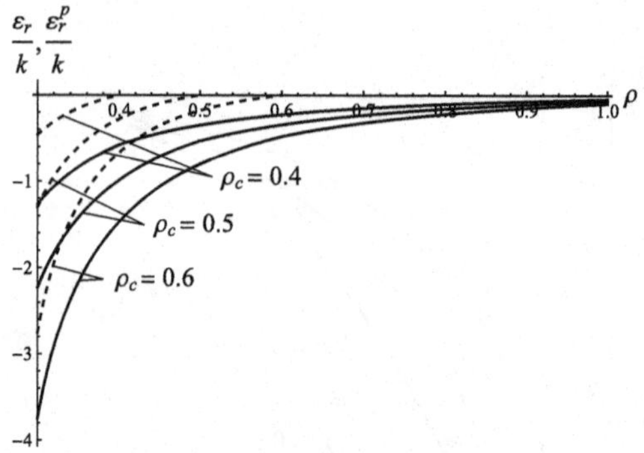

Fig. 2.17 Variation of the total and plastic radial strains in an $a = 0.3$ disc at $\eta = 0.82$, $\eta_1 = 0.735$ and several values of ρ_c

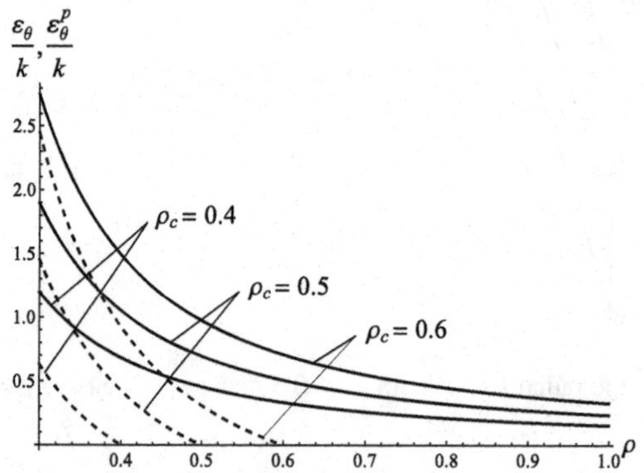

Fig. 2.18 Variation of the total and plastic circumferential strains in an $a = 0.3$ disc at $\eta = 0.82$, $\eta_1 = 0.735$ and several values of ρ_c

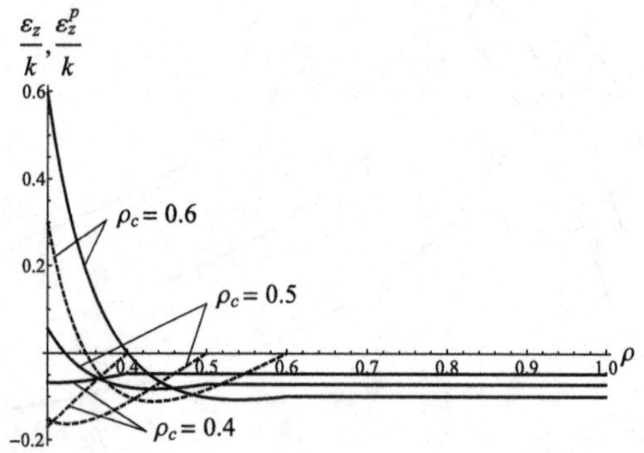

Fig. 2.19 Variation of the total and plastic axial strains in an $a = 0.3$ disc at $\eta = 0.82$, $\eta_1 = 0.735$ and several values of ρ_c

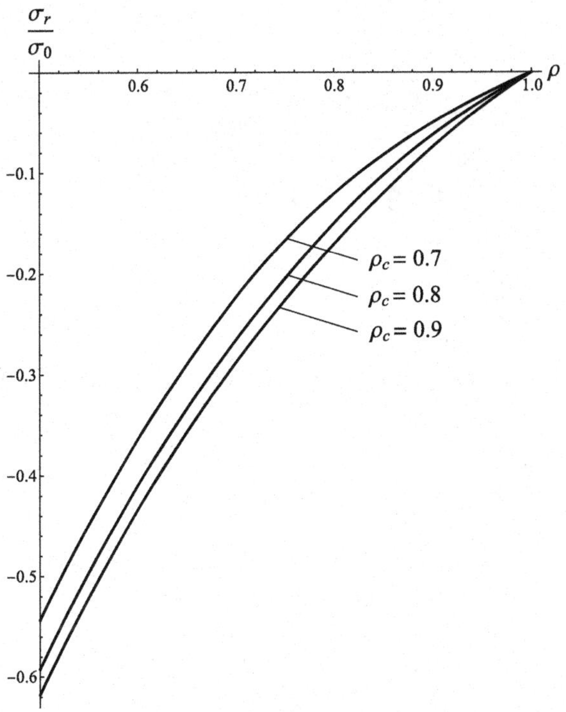

Fig. 2.20 Variation of the radial stress in an $a = 0.5$ disc at $\eta = 0.82$, $\eta_1 = 0.735$ and several values of ρ_c

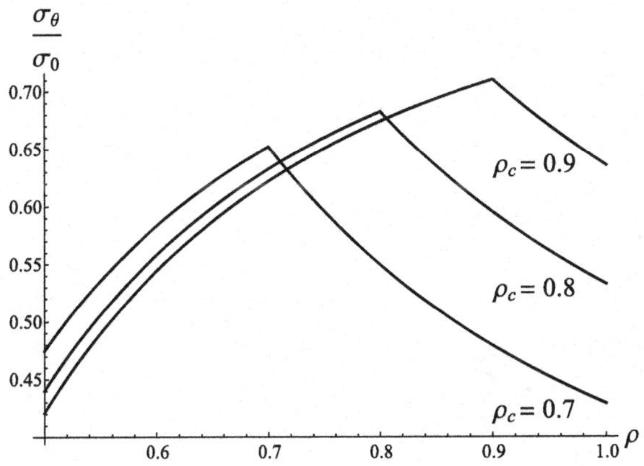

Fig. 2.21 Variation of the circumferential stress in an $a = 0.5$ disc at $\eta = 0.82$, $\eta_1 = 0.735$ and several values of ρ_c

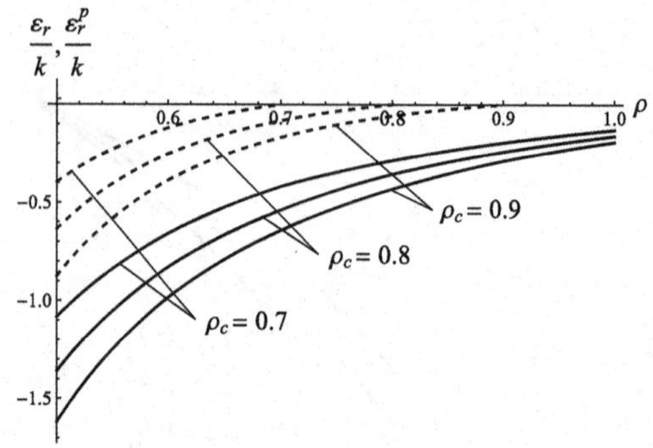

Fig. 2.22 Variation of the total and plastic radial strains in an $a = 0.5$ disc at $\eta = 0.82$, $\eta_1 = 0.735$ and several values of ρ_c

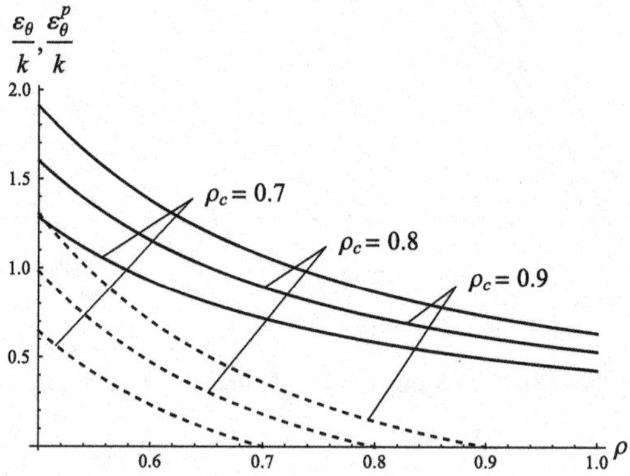

Fig. 2.23 Variation of the total and plastic circumferential strains in an $a = 0.5$ disc at $\eta = 0.82$, $\eta_1 = 0.735$ and several values of ρ_c

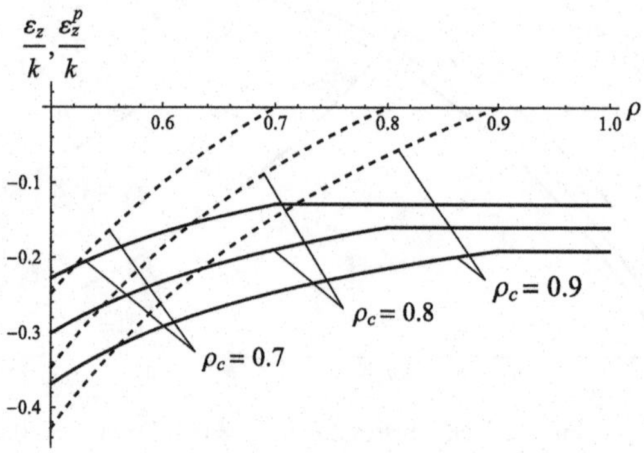

Fig. 2.24 Variation of the total and plastic axial strains in an $a = 0.5$ disc at $\eta = 0.82$, $\eta_1 = 0.735$ and several values of ρ_c

Since the plastic zone starts to develop from the inner radius of the disc, the material is elastic in the range $\rho_c \le \rho \le 1$. Therefore, Eq. (1.29) are valid in this range. However, A and B are not determined by Eq. (2.4). Nevertheless, the radial stress determined by Eq. (1.29) must satisfy the boundary condition (2.1). Then,

$$A + B = 0. \tag{2.81}$$

The radial stress determined by Eq. (1.53) must satisfy the boundary condition (2.2). Therefore,

$$3\beta_0 - \frac{\beta_1}{2}\left(1 + 3\sqrt{3}\beta_1\right)\sin\psi_a + \frac{\sqrt{3}}{2}\beta_1\left(1 - \sqrt{3}\beta_1\right)\cos\psi_a = -q \tag{2.82}$$

where ψ_a is the value of ψ at $\rho = a$. It is evident that $\sigma_\theta > 0$ at $\rho = a$. It is seen from this inequality and Eq. (1.53) that

$$3\beta_0 + \frac{\beta_1}{2}\left(1 - 3\sqrt{3}\beta_1\right)\sin\psi_a - \frac{\sqrt{3}}{2}\beta_1\left(1 + \sqrt{3}\beta_1\right)\cos\psi_a > 0.$$

This inequality and Eq. (2.82) allow a unique value of ψ_a to be determined. It follows from Eqs. (1.27), (1.29) and (1.53) that

$$3\beta_0 - \frac{\beta_1}{2}\left(1 + 3\sqrt{3}\beta_1\right)\sin\psi_c + \frac{\sqrt{3}}{2}\beta_1\left(1 - \sqrt{3}\beta_1\right)\cos\psi_c = -A\left(1 - \frac{1}{\rho_c^2}\right), \tag{2.83}$$

$$3\beta_0 + \frac{\beta_1}{2}\left(1 - 3\sqrt{3}\beta_1\right)\sin\psi_c - \frac{\sqrt{3}}{2}\beta_1\left(1 + \sqrt{3}\beta_1\right)\cos\psi_c = -A\left(1 + \frac{1}{\rho_c^2}\right).$$

Here B has been eliminated by means of Eq. (2.81). It is convenient to put $\psi_0 = \psi_a = p$ and $\rho_0 = a$ in Eqs. (1.16) and (1.57). Then,

$$\rho = a\exp\left[\frac{3\beta_1}{2}(\psi - \psi_a)\right]\sqrt{\frac{\sin(\psi_a - \pi/3)}{\sin(\psi - \pi/3)}}. \tag{2.84}$$

It follows from this equation and the definition for ψ_c that

$$\rho_c = a\exp\left[\frac{3\beta_1}{2}(\psi_c - \psi_a)\right]\sqrt{\frac{\sin(\psi_a - \pi/3)}{\sin(\psi_c - \pi/3)}}. \tag{2.85}$$

Solving Eq. (2.83) for ρ_c and A gives

$$\rho_c^2 = \frac{3\beta_0 - 3\beta_1^2\sin(\psi_c + \pi/6)}{\beta_1\sin(\psi_c - \pi/3)}, \quad A = 3\beta_1^2\sin\left(\psi_c + \frac{\pi}{6}\right) - 3\beta_0. \tag{2.86}$$

The plastic zone starts to develop at $q = q_e$, $\rho_c = a$ and $\psi_a = \psi_c = \psi_e$. Then, it follows from Eqs. (2.80), (2.82) and (2.86) that

$$3\beta_0 - \frac{\beta_1}{2}\left(1+3\sqrt{3}\beta_1\right)\sin\psi_e + \frac{\sqrt{3}}{2}\beta_1\left(1-\sqrt{3}\beta_1\right)\cos\psi_e + \\ + \frac{3\left(1-a^2\right)\left(3\sqrt{3+a^4}-2\alpha a^2\right)}{27+a^4\left(9-4\alpha^2\right)} = 0, \quad a^2 = \frac{3\beta_0 - 3\beta_1^2\sin\left(\psi_e+\pi/6\right)}{\beta_1\sin\left(\psi_e-\pi/3\right)}. \tag{2.87}$$

These equations are compatible if the value of ψ_e is given by

$$\psi_e = 2\arctan\left[\frac{3\beta_1^2+\sqrt{9\beta_1^4+\beta_1^2a^4-9\beta_0^2}}{3\beta_0-\beta_1 a^2}\right] + \frac{11\pi}{6}. \tag{2.88}$$

Thus the value of ψ_e is in the range

$$\frac{11\pi}{6} - 2\operatorname{arccot}\left(\frac{2\alpha\sqrt{9-4\alpha^2}}{1+\sqrt{9-4\alpha^2}}\right) \le \psi_e < \frac{11\pi}{6} - 2\arctan\left(\frac{3\sqrt{3}+2\sqrt{9-4\alpha^2}}{2\sqrt{3}\alpha+\sqrt{9-4\alpha^2}}\right) \tag{2.89}$$

when a varies in the range $0 \le a < 1$. Here Eq. (1.54) has been used. Differentiating Eqs. (2.82) and (2.86) for ρ_c with respect to q and, then, putting $\psi_a = \psi_e$ and $\psi_c = \psi_e$ yield the values of the derivative $d\psi_a/dq$ and $d\rho_c^2/dq$ at $\psi_c = \psi_a = \psi_e$ in the form

$$\frac{d\psi_a}{dq} = \frac{2}{\beta_1\left[\left(1+3\sqrt{3}\beta_1\right)\cos\psi_e + \sqrt{3}\left(1-\sqrt{3}\beta_1\right)\sin\psi_e\right]}, \tag{2.90}$$

$$\frac{d\rho_c^2}{dq} = \frac{6\left(2\beta_1^2-\beta_0\cos\psi_e-\sqrt{3}\beta_0\sin\psi_e\right)}{\beta_1\left(\sin\psi_e-\sqrt{3}\cos\psi_e\right)^2}\frac{d\psi_c}{dq}.$$

It is possible to verify by inspection that the right hand side of Eq. $(2.90)^1$ is negative and the right hand side of Eq. $(2.90)^2$ is positive in the range (2.89) for typical values of α used in the constitutive equations for metals [8–10]. Therefore, $d\psi_a/dq < 0$ and $d\psi_c/dq > 0$ at $\psi_a = \psi_e$. Hence it is possible to conclude that the value of ψ varies in the range

$$\psi_a \le \psi \le \psi_c. \tag{2.91}$$

The entire disc becomes plastic when $\rho_c = 1$. The corresponding value of ψ_c is denoted by ψ_q. It follows from Eq. (1.53) that the equation for ψ_q is

$$3\beta_0 - \frac{\beta_1}{2}\left(1 + 3\sqrt{3}\beta_1\right)\sin\psi_q + \frac{\sqrt{3}}{2}\beta_1\left(1 - \sqrt{3}\beta_1\right)\cos\psi_q = 0.$$

The solution of this equation compatible with the inequality $\sigma_\theta > 0$ where σ_θ is given by (1.53) is

$$\psi_q = 2\arctan\left[\frac{\beta_1\left(1 + 3\sqrt{3}\beta_1\right) + 2\sqrt{\beta_1^2\left(1 + 9\beta_1^2\right) - 9\beta_0^2}}{6\beta_0 - \sqrt{3}\beta_1\left(1 - \sqrt{3}\beta_1\right)}\right] + 2\pi. \qquad (2.92)$$

Substituting Eq. (2.85) into Eq. (2.86) leads to the following equation that connects ψ_a and ψ_c

$$a^2 = \frac{3\left[\beta_0 - \beta_1^2\sin\left(\psi_c + \pi/6\right)\right]}{\beta_1\sin\left(\psi_a - \pi/3\right)}\exp\left[3\beta_1\left(\psi_a - \psi_c\right)\right]. \qquad (2.93)$$

It follows from Eq. (2.82) that

$$\frac{dq}{d\psi_a} = \frac{\beta_1}{2}\left[\left(1 + 3\sqrt{3}\beta_1\right)\cos\psi_a + \sqrt{3}\left(1 - \sqrt{3}\beta_1\right)\sin\psi_a\right].$$

It is seen from this equation that $dq/d\psi_a = 0$ at

$$\psi_a = \psi_l = \arctan\left[\frac{1 + 3\sqrt{3}\beta_1}{\sqrt{3}\left(\sqrt{3}\beta_1 - 1\right)}\right]. \qquad (2.94)$$

The corresponding value of q is denoted by q_m. The value of q_m is found from Eqs. (2.82) and (2.94). The plane stress approximation is not valid in the range $q > q_m$ since an intensive local thickening occurs in the vicinity of $\rho = a$. Therefore, the present solution is restricted to the range $\psi_e \geq \psi_a \geq \psi_l$. Another restriction follows from Eq. (2.92) and is $\psi_e \leq \psi_c \leq \psi_q$. Putting $\psi_a = \psi_l$ and $\psi_c = \psi_q$ in Eq. (2.93) gives the following equation for a_{cr}

$$a_{cr}^2 = \frac{3\left[\beta_0 - \beta_1^2\sin\left(\psi_q + \pi/6\right)\right]}{\beta_1\sin\left(\psi_l - \pi/3\right)}\exp\left[3\beta_1\left(\psi_l - \psi_q\right)\right]. \qquad (2.95)$$

Substituting Eqs. (1.54), (2.92) and (2.94) into Eq. (2.95) supplies the dependence of a_{cr} on α. This dependence is illustrated in Fig. 2.25. The present solution is restricted

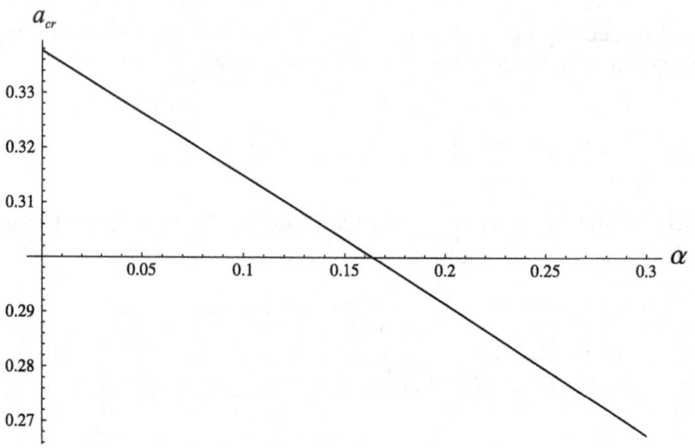

Fig. 2.25 Variation of a_{cr} with α

by the inequality $\psi_e \geq \psi_a \geq \psi_l$ if $a \leq a_{cr}$ and by the inequality $\psi_e \leq \psi_c \leq \psi_q$ if $a \geq a_{cr}$. Replacing ψ_0 with ψ_a in Eq. (1.91) and using Eq. (1.87) lead to

$$\frac{\left[\left(1+3\sqrt{3}\beta_1\right)\cos\psi + \sqrt{3}\left(1-\sqrt{3}\beta_1\right)\sin\psi\right]}{\left(\sqrt{3}\cos\psi - \sin\psi\right)}d\psi =$$
$$= \frac{\left[\left(1+3\sqrt{3}\beta_1\right)\cos\psi_a + \sqrt{3}\left(1-\sqrt{3}\beta_1\right)\sin\psi_a\right]}{\left(\sqrt{3}\cos\psi_a - \sin\psi_a\right)}d\psi_a$$

Integrating this equation gives

$$\frac{\sin(\psi_a - \pi/3)}{\sin(\psi - \pi/3)}\exp\left[3\beta_1(\psi - \psi_a)\right] = \frac{C^2}{a^2} \tag{2.96}$$

where C is constant on each characteristic curve. Comparing Eqs. (2.84) and (2.96) shows that $C = \rho$. Therefore, integrating Eq. (1.92) is equivalent to integrating at a fixed value of ρ. Since $p = \psi_a$, it is possible to rewrite Eq. (1.92) as

$$\frac{d\varepsilon_r}{d\psi_a} = \xi_r, \quad \frac{d\varepsilon_\theta}{d\psi_a} = \xi_\theta, \quad \frac{d\varepsilon_z}{d\psi_a} = \xi_z. \tag{2.97}$$

Differentiating Eqs. (2.81) and (2.86) for A with respect to ψ_a gives

$$\frac{dA}{d\psi_a} = 3\beta_1^2\cos\left(\psi_c + \frac{\pi}{6}\right)\frac{d\psi_c}{d\psi_a}, \quad \frac{dB}{d\psi_a} = -\frac{dA}{d\psi_a}. \tag{2.98}$$

The derivative $d\psi_c/d\psi_a$ is found from Eq. (2.93) as

$$\frac{d\psi_c}{d\psi_a} = \frac{a^2\left[3\beta_1 \sin\left(\psi_a - \pi/3\right) - \cos\left(\psi_a - \pi/3\right)\right]}{3\beta_1\left\{\cos\left(\psi_c + \pi/6\right)\exp\left[3\beta_1\left(\psi_a - \psi_c\right)\right] + a^2 \sin\left(\psi_a - \pi/3\right)\right\}}. \tag{2.99}$$

Using Eq. (2.86) for ρ_c and Eq. (2.98) it is possible to transform Eq. (1.63) to

$$\frac{\xi_c}{k} = -\left[\frac{(1+\nu)\,\beta_1 \sin\left(\psi_c - \pi/3\right)}{\beta_0 - \beta_1^2 \sin\left(\psi_c + \pi/6\right)} + 3\,(1-\nu)\right]\beta_1^2 \cos\left(\psi_c + \frac{\pi}{6}\right)\frac{d\psi_c}{d\psi_a}. \tag{2.100}$$

Equations (1.90) and (1.96) become

$$\frac{\xi_\theta}{k} = \left\{\Omega_P\left(\psi_a\right)\int\limits_{\psi_c}^{\psi}\exp\left[-\int\limits_{\psi_c}^{\mu_1} W_0(\mu)\,d\mu\right] W_1\left(\mu_1\right) d\mu_1 + \frac{\xi_c}{k}\right\}\times$$
$$\times \exp\left[\int\limits_{\psi_c}^{\psi} W_0(\mu)\,d\mu\right], \tag{2.101}$$

$$\frac{\xi_\theta}{k} = \Omega_P\left(\psi_a\right)\exp\left[3\beta_1\left(\psi_c - \psi\right)\right]\int\limits_{\psi_c}^{\psi}\exp\left[3\beta_1\left(\mu - \psi_c\right)\right] W_1(\mu)\,d\mu +$$
$$+ \frac{\xi_c}{k}\exp\left[3\beta_1\left(\psi_c - \psi\right)\right],$$

respectively. It has been taken into account here that $d\psi_0/dp = 1$ in the case under consideration. Equation $(2.101)^1$ is valid for plastically incompressible materials and Eq. $(2.101)^2$ for plastically compressible materials. In these equations $\Omega_P\left(\psi_a\right)$ should be eliminated by means of Eq. (1.87). Using Eq. (2.99) the right hand side of Eq. (2.100) is represented as a function of ψ_a and ψ_c. Therefore, the right hand side of each of Eq. (2.101) depends on ψ, ψ_a and ψ_c. Eliminating ψ_c by means of the solution of Eq. (2.93) supplies the right hand sides of these equations as functions of ψ and ψ_a. Further eliminating ψ by means of the solution of Eq. (2.96) at any given value of $C = \rho$ determines the right hand side of Eq. (2.97) for ε_θ as a function of ψ_a. This function is denoted by $E_\theta^{(i)}\left(\psi_a\right)$ for plastically incompressible materials and $E_\theta^{(c)}\left(\psi_a\right)$ for plastically compressible materials. Therefore, Eq. (2.97) for ε_θ can be integrated numerically. In particular, the value of ε_θ at $\psi_a = \psi_m$ and $\rho = C$ is given by

$$\varepsilon_\theta = \int\limits_{\psi_i}^{\psi_m} E_\theta^{(i)}\left(\psi_a\right) d\psi_a + E_\theta^e, \quad \varepsilon_\theta = \int\limits_{\psi_i}^{\psi_m} E_\theta^{(c)}\left(\psi_a\right) d\psi_a + E_\theta^e \tag{2.102}$$

for plastically incompressible and plastically compressible materials, respectively. Here ψ_m is prescribed and Eq. (2.102) supplies ε_θ in the plastic zone. It is seen from Eq. (2.82) that prescribing the value of ψ_m is equivalent to prescribing the value of q. The procedure to find ψ_i and E_θ^e is as follows. The value of E_θ^e is the circumferential strain at $\rho = \rho_c = C$ and the value of ψ_i is the value of ψ_a at $\rho_c = C$. It is seen from Eq. (1.26)[1] that E_θ^e is determined from the solution in the elastic zone. Alternatively, since the stresses are continuous across the elastic/plastic boundary, the value of E_θ^e can be found from Eqs. (1.1) and (1.53). Let ψ_C be the value of ψ_c at $\rho_c = C$. Then, the equation for ψ_C follows from Eq. (2.86) as

$$C^2 = \frac{3\beta_0 - 3\beta_1^2 \sin(\psi_C + \pi/6)}{\beta_1 \sin(\psi_C - \pi/3)}. \tag{2.103}$$

Having found the value of ψ_C the value of E_θ^e is determined from Eqs. (1.1) and (1.53) at $\psi = \psi_C$ as

$$\frac{E_\theta^e}{k} = (1+\nu)\beta_1 \sin\left(\psi_C - \frac{\pi}{3}\right) + 3(1-\nu)\left[\beta_0 - \beta_1^2 \sin\left(\psi_C + \frac{\pi}{6}\right)\right]. \tag{2.104}$$

Substituting $\psi_c = \psi_C$ into Eq. (2.93) and solving this equation for ψ_a supplies the value of ψ_i involved in Eq. (2.102).

The distributions of ε_r and ε_z in the plastic zone are determined in a similar manner. In particular, using Eqs. (1.22), (1.86) and (1.93)

$$\begin{aligned}
\xi_r &= \xi_r^e + \xi_r^p = \xi_r^e + \xi_\theta^p \left[\frac{\beta_1\left(\sqrt{3} - \beta_1\right)\cos\psi - \beta_1\left(1 + \sqrt{3}\beta_1\right)\sin\psi + 2\beta_0}{\beta_1\left(1 - \sqrt{3}\beta_1\right)\sin\psi - \beta_1\left(\sqrt{3} + \beta_1\right)\cos\psi + 2\beta_0}\right] = \\
&= \xi_r^e + \left(\xi_\theta - \xi_\theta^e\right)\left[\frac{\beta_1\left(\sqrt{3} - \beta_1\right)\cos\psi - \beta_1\left(1 + \sqrt{3}\beta_1\right)\sin\psi + 2\beta_0}{\beta_1\left(1 - \sqrt{3}\beta_1\right)\sin\psi - \beta_1\left(\sqrt{3} + \beta_1\right)\cos\psi + 2\beta_0}\right], \\
\xi_z &= \xi_z^e + \xi_z^p = \xi_z^e + 2\xi_\theta^p \left[\frac{\beta_1^2\left(\cos\psi + \sqrt{3}\sin\psi\right) - 2\beta_0}{\beta_1\left(1 - \sqrt{3}\beta_1\right)\sin\psi - \beta_1\left(\sqrt{3} + \beta_1\right)\cos\psi + 2\beta_0}\right] = \\
&= \xi_z^e + 2\left(\xi_\theta - \xi_\theta^e\right)\left[\frac{\beta_1^2\left(\cos\psi + \sqrt{3}\sin\psi\right) - 2\beta_0}{\beta_1\left(1 - \sqrt{3}\beta_1\right)\sin\psi - \beta_1\left(\sqrt{3} + \beta_1\right)\cos\psi + 2\beta_0}\right].
\end{aligned} \tag{2.105}$$

for plastically incompressible materials and

$$\begin{aligned}
\xi_r &= \xi_r^e + \xi_r^p = \xi_r^e + \xi_\theta^p \left[\frac{3\beta_1 \sin(\psi - \pi/3) + \sin(\psi + \pi/6)}{\sin(\psi + \pi/6) - 3\beta_1 \sin(\psi - \pi/3)}\right] = \\
&= \xi_r^e + \left(\xi_\theta - \xi_\theta^e\right)\left[\frac{3\beta_1 \sin(\psi - \pi/3) + \sin(\psi + \pi/6)}{\sin(\psi + \pi/6) - 3\beta_1 \sin(\psi - \pi/3)}\right],
\end{aligned} \tag{2.106}$$

$$\xi_z = \xi_z^e + \xi_z^p = \xi_z^e - \frac{2\beta_1^2 \xi_\theta^p}{3} \left[\frac{9\alpha + (9 + 2\alpha^2) \sin(\psi + \pi/6)}{\sin(\psi + \pi/6) - 3\beta_1 \sin(\psi - \pi/3)} \right] =$$

$$= \xi_z^e - \frac{2\beta_1^2 (\xi_\theta - \xi_\theta^e)}{3} \left[\frac{9\alpha + (9 + 2\alpha^2) \sin(\psi + \pi/6)}{\sin(\psi + \pi/6) - 3\beta_1 \sin(\psi - \pi/3)} \right].$$

for plastically compressible materials. Using Eqs. (1.85), (1.87) and (2.101) the right hand sides of Eqs. (2.105) and (2.106) are expressed in terms of ψ, ψ_a and ψ_c. Then, ψ_c can be eliminated by means of the solution to Eq. (2.93). Eliminating ψ by means of the solution to Eq. (2.96) at a given value of C determines the right hand sides of these equations as functions of ψ_a. These functions are denoted by $E_r^{(i)}(\psi_a)$ and $E_z^{(i)}(\psi_a)$ for plastically incompressible materials and $E_r^{(c)}(\psi_a)$ and $E_z^{(c)}(\psi_a)$ for plastically compressible materials. Therefore, Eq. (2.97) for ε_r and ε_z can be integrated numerically. In particular,

$$\varepsilon_r = \int_{\psi_i}^{\psi_m} E_r^{(i)}(\psi_a)\, d\psi_a + E_r^e, \quad \varepsilon_z = \int_{\psi_i}^{\psi_m} E_z^{(i)}(\psi_a)\, d\psi_a + E_z^e \tag{2.107}$$

for plastically incompressible materials and

$$\varepsilon_r = \int_{\psi_i}^{\psi_m} E_r^{(c)}(\psi_a)\, d\psi_a + E_r^e, \quad \varepsilon_z = \int_{\psi_i}^{\psi_m} E_z^{(c)}(\psi_a)\, d\psi_a + E_z^e \tag{2.108}$$

for plastically compressible materials. Here E_r^e and E_z^e are the radial and axial strains, respectively, at $\rho = \rho_c = C$. These strains are determined from Eq. (1.61) at $\rho = \rho_c = C$, (2.81) and (2.86) or from Eqs. (1.1) and (1.53) at $\psi = \psi_C$. As a result,

$$\frac{E_r^e}{k} = -(1 + \nu)\beta_1 \sin\left(\psi_C - \frac{\pi}{3}\right) + 3(1 - \nu)\left[\beta_0 - \beta_1^2 \sin\left(\psi_C + \frac{\pi}{6}\right)\right], \tag{2.109}$$

$$\frac{E_z^e}{k} = 6\nu\left[\beta_1^2 \sin\left(\psi_C + \frac{\pi}{6}\right) - \beta_0\right]$$

for both incompressible and compressible materials. Having found the distributions of the total strains in the plastic zone it is possible to determine their plastic portion by means of Eq. (1.3) in which the elastic strains should be eliminated using Eqs. (1.1), (1.53) and (2.84). The total strains in the elastic zone follow from Eq. (1.61) in which A and B should be eliminated by means of Eq. (2.81) and (2.86). The value of ψ_c involved in Eq. (2.86) is determined from Eq. (2.93) assuming that $\psi_a = \psi_m$.

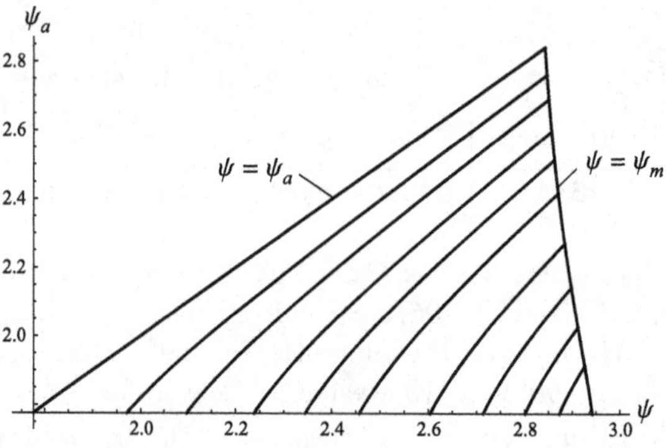

Fig. 2.26 Field of characteristics for an $a = 0.2$ disc at $\alpha = 0.3$

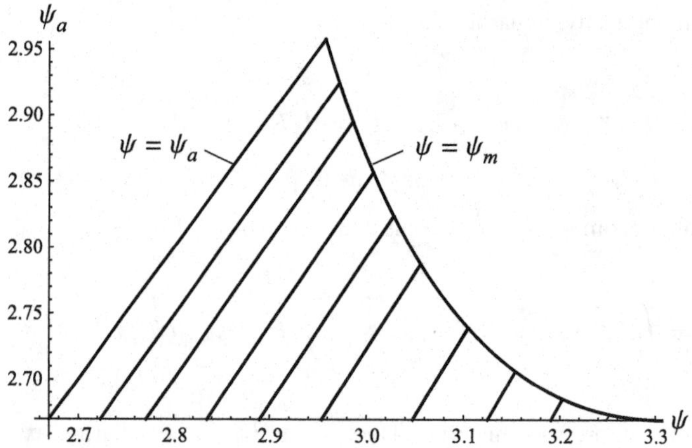

Fig. 2.27 Field of characteristics for an $a = 0.5$ disc at $\alpha = 0.3$

The solution found is illustrated for $\alpha = 0.3$. This value of α has been reported in [10]. Two fields of characteristics found from Eq. (2.96) are shown in Fig. 2.26 for $a = 0.2 < a_{cr}$ and in Fig. 2.27 for $a = 0.5 > a_{cr}$. The distributions of the radial and circumferential stresses are depicted in Figs. 2.28 and 2.29 for an $a = 0.2 < a_{cr}$ disc and in Figs. 2.30 and 2.31 for an $a = 0.5 > a_{cr}$ disc for several values of ρ_c. The distributions of the radial, circumferential and axial strains for plastically incompressible material are shown in Figs. 2.32, 2.33 and 2.34 for an $a = 0.2$ disc and in Figs. 2.35, 2.36 and 2.37 for an $a = 0.5$ disc. The distributions of these strains for plastically compressible material are shown in Figs. 2.38, 2.39 and 2.40 for an $a = 0.2$ disc and in Figs. 2.41, 2.42 and 2.43 for an $a = 0.5$ disc. The solid lines correspond to the total strains and the broken lines to the plastic strains. In all calculations $\nu = 0.3$.

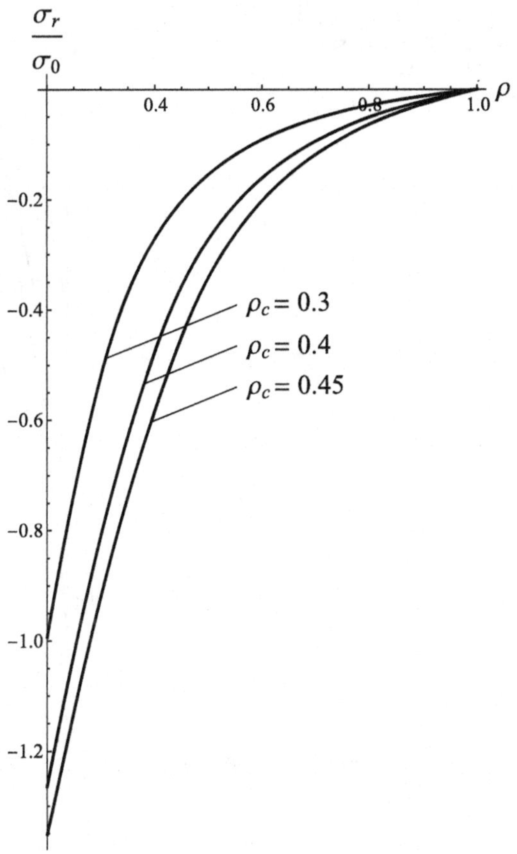

Fig. 2.28 Variation of the radial stress in an $a = 0.2$ disc at $\alpha = 0.3$ and several values of ρ_c

2.2 Disc Under External Pressure

The disc shown in Fig. 1.1 is loaded by external pressure q_0 and its inner surface is stress free. Therefore, $\tau = 0$, $d\tau/dp = 0$, $\varepsilon_r^T = \varepsilon_\theta^T = \varepsilon_z^T = 0$, and $\xi_r^T = \xi_\theta^T = \xi_z^T = 0$. The boundary conditions are

$$\sigma_r = -q_0 \quad \text{for} \quad \rho = 1, \tag{2.110}$$

$$\sigma_r = 0 \quad \text{for} \quad \rho = a. \tag{2.111}$$

At the stage of purely elastic loading these boundary conditions and Eq. (1.29) combine to give

$$A + B = -q, \quad A + a^2 B = 0 \tag{2.112}$$

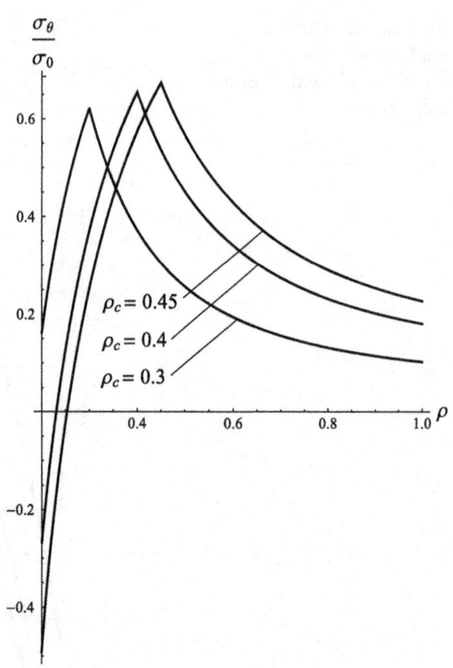

Fig. 2.29 Variation of the circumferential stress in an $a = 0.2$ disc at $\alpha = 0.3$ and several values of ρ_c

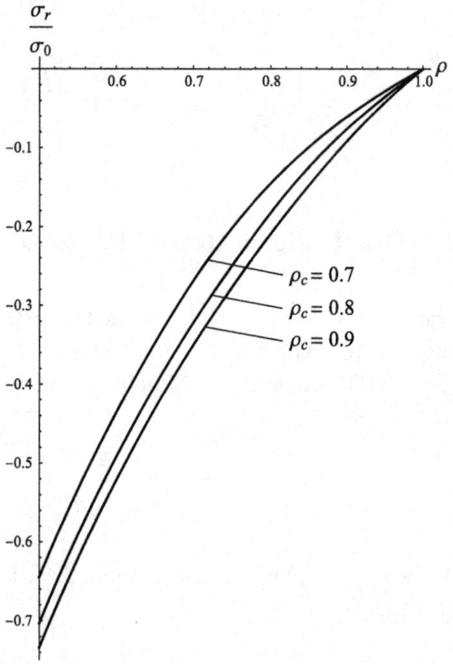

Fig. 2.30 Variation of the radial stress in an $a = 0.5$ disc at $\alpha = 0.3$ and several values of ρ_c

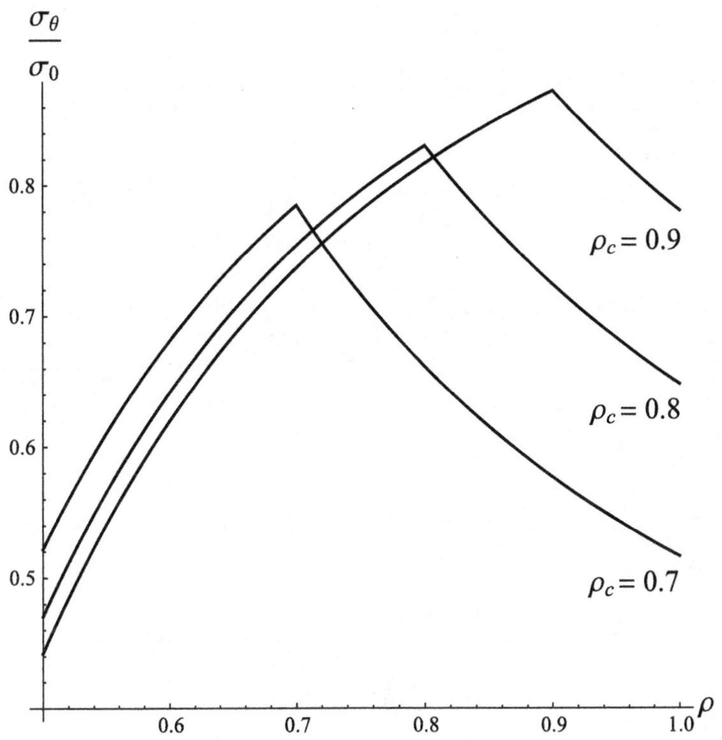

Fig. 2.31 Variation of the circumferential stress in an $a = 0.5$ disc at $\alpha = 0.3$ and several values of ρ_c

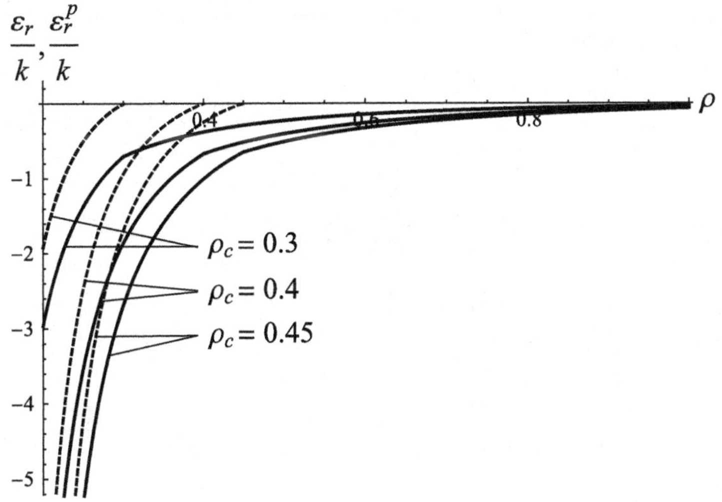

Fig. 2.32 Variation of the total and plastic radial strains in an $a = 0.2$ disc of plastically incompressible material at $\alpha = 0.3$ and several values of ρ_c

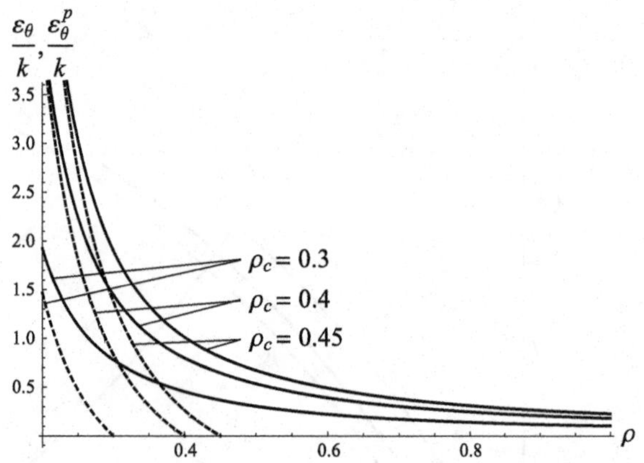

Fig. 2.33 Variation of the total and plastic circumferential strains in an $a = 0.2$ disc of plastically incompressible material at $\alpha = 0.3$ and several values of ρ_c

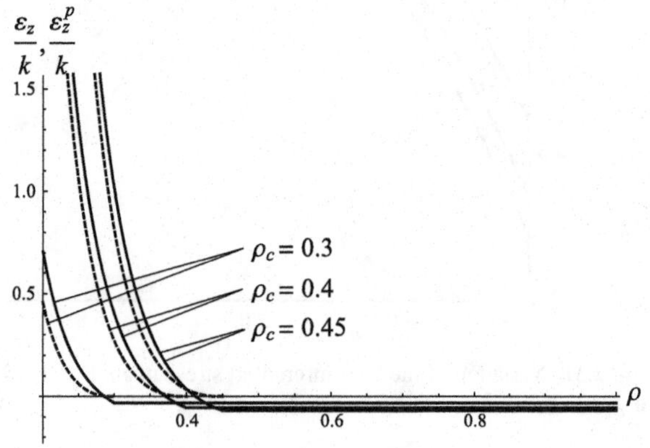

Fig. 2.34 Variation of the total and plastic axial strains in an $a = 0.2$ disc of plastically incompressible material at $\alpha = 0.3$ and several values of ρ_c

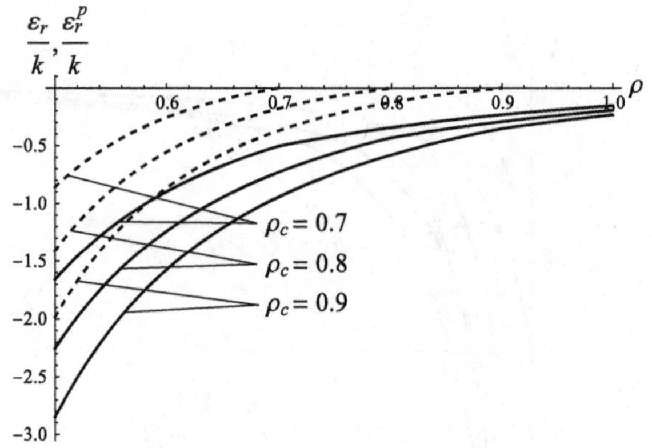

Fig. 2.35 Variation of the total and plastic radial strains in an $a = 0.5$ disc of plastically incompressible material at $\alpha = 0.3$ and several values of ρ_c

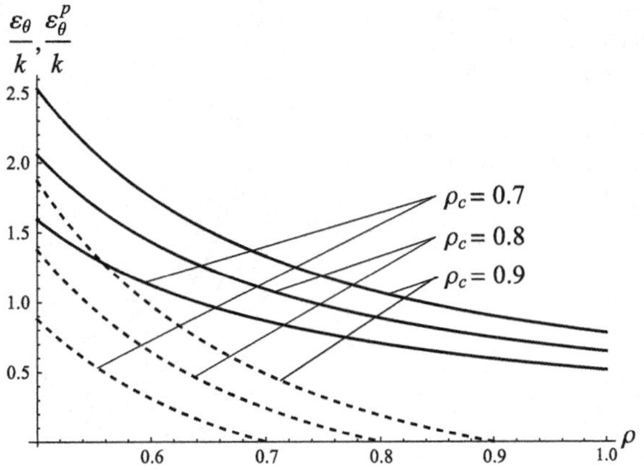

Fig. 2.36 Variation of the total and plastic circumferential strains in an $a = 0.5$ disc of plastically incompressible material at $\alpha = 0.3$ and several values of ρ_c

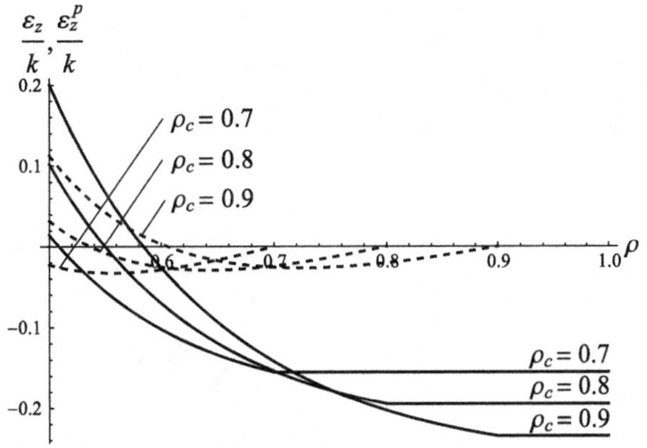

Fig. 2.37 Variation of the total and plastic axial strains of plastically incompressible material in an $a = 0.5$ disc at $\alpha = 0.3$ and several values of ρ_c

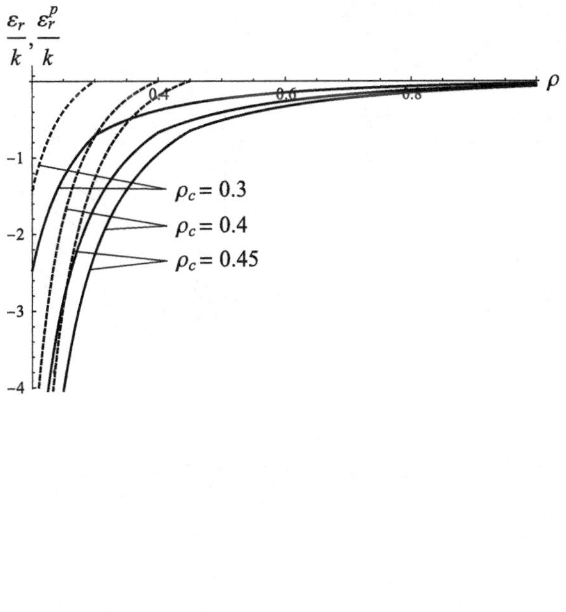

Fig. 2.38 Variation of the total and plastic radial strains in an $a = 0.2$ disc of plastically compressible material at $\alpha = 0.3$ and several values of ρ_c

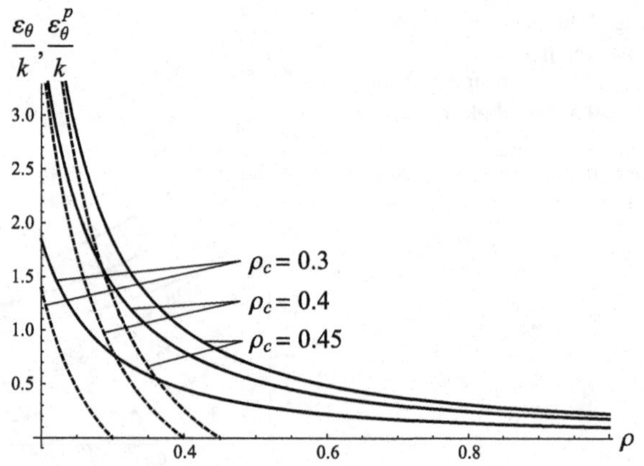

Fig. 2.39 Variation of the total and plastic circumferential strains in an $a = 0.2$ disc of plastically compressible material at $\alpha = 0.3$ and several values of ρ_c

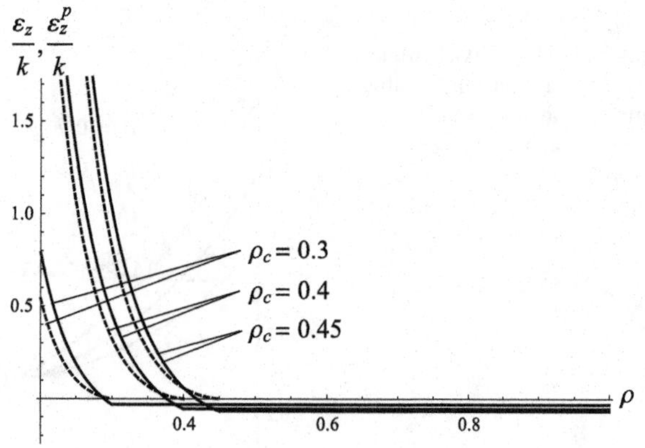

Fig. 2.40 Variation of the total and plastic axial strains in an $a = 0.2$ disc of plastically compressible material at $\alpha = 0.3$ and several values of ρ_c

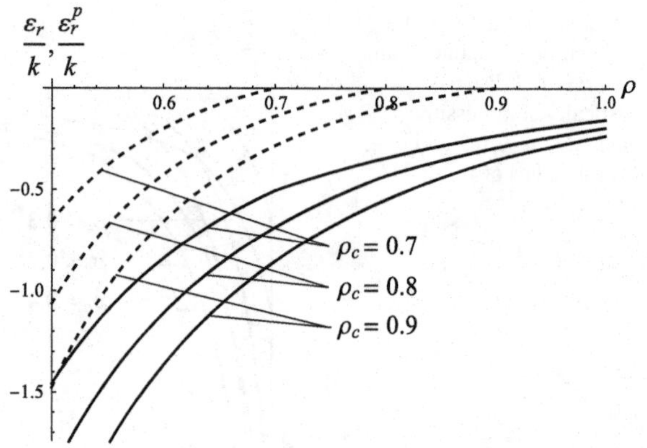

Fig. 2.41 Variation of the total and plastic radial strains in an $a = 0.5$ disc of plastically compressible material at $\alpha = 0.3$ and several values of ρ_c

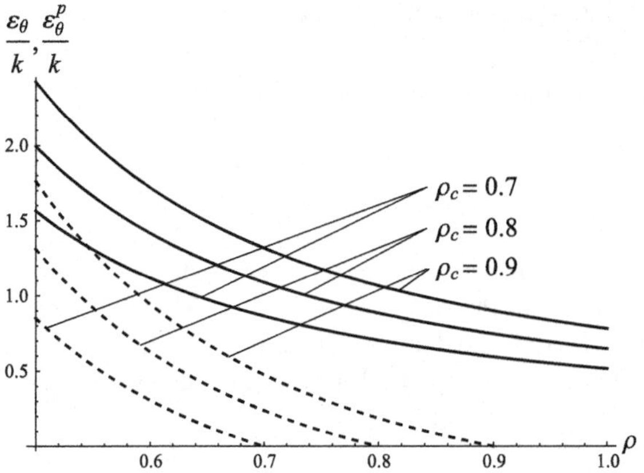

Fig. 2.42 Variation of the total and plastic circumferential strains in an $a = 0.5$ disc of plastically compressible material at $\alpha = 0.3$ and several values of ρ_c

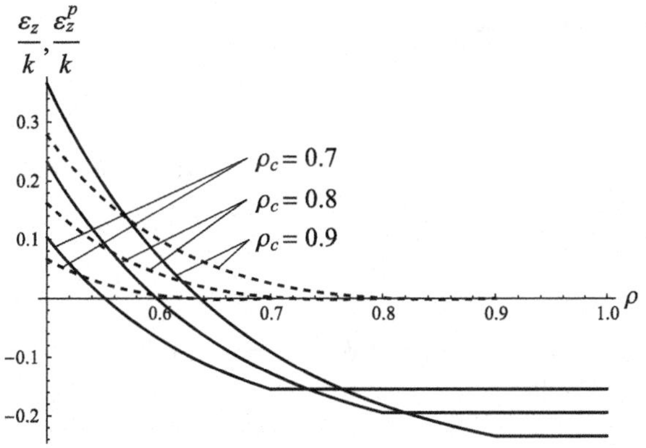

Fig. 2.43 Variation of the total and plastic axial strains of plastically compressible material in an $a = 0.5$ disc at $\alpha = 0.3$ and several values of ρ_c

where $q = q_0/\sigma_0$. Solving Eq. (2.112) for A and B yields

$$A = \frac{a^2 q}{1 - a^2}, \quad B = -\frac{q}{1 - a^2}. \tag{2.113}$$

This equation and one of the yield criteria combine to determine the value of q corresponding to the initiation of plastic yielding. This value of q will be denoted by q_e. The corresponding value of the function ψ involved in Eqs. (1.32), (1.47) and (1.53) will be denoted by ψ_e. It is understood here that ψ is calculated at the site of plastic yielding initiation and the plastic zone reduces to a circle at this instant. The solutions considered in this chapter are for elastic/plastic discs. Therefore, it is assumed that $q \geq q_e$ and there is an elastic/plastic boundary, $\rho = \rho_c$, where $\psi = \psi_c$.

2.2.1 Yield Criterion (1.5)

Substituting Eq. (2.113) into Eq. (1.31) results in

$$q_e = \frac{1 - a^2}{2}. \tag{2.114}$$

Since the plastic zone starts to develop from the inner radius of the disc, the material is elastic in the range $\rho_c \le \rho \le 1$. Therefore, Eq. (1.29) are valid in this range. However, A and B are not determined by Eq. (2.113). Nevertheless, the radial stress determined by Eq. (1.29) must satisfy the boundary condition (2.110). Then,

$$A + B = -q. \tag{2.115}$$

The radial stress determined by Eq. (1.32) must satisfy the boundary condition (2.111). Hence,

$$\sin \psi_a = 0. \tag{2.116}$$

It is evident that $\sigma_\theta < 0$ at $\rho = a$. Therefore, it follows from the solution (1.32) for σ_θ and Eq. (2.116) that

$$\psi_a = 0. \tag{2.117}$$

Here ψ_a is the value of ψ at $\rho = a$. It follows from Eqs. (1.27), (1.29), (1.32) and (2.115) that

$$A\left(1 - \frac{1}{\rho_c^2}\right) + q = \frac{2}{\sqrt{3}} \sin \psi_c, \quad A\left(\frac{1}{\rho_c^2} + 1\right) + q = \cos \psi_c + \frac{\sin \psi_c}{\sqrt{3}}. \tag{2.118}$$

It is convenient to put $\psi_0 = \psi_a = 0$ and $\rho_0 = a$ in Eq. (1.35). Then, this equation becomes

$$\rho = \frac{a\sqrt{\sqrt{3}}}{\sqrt{2}\sqrt{\sin(\pi/3 - \psi)}} \exp\left(\frac{\sqrt{3}}{2}\psi\right). \tag{2.119}$$

Using this equation and the definition for ψ_c the radius of the elastic/plastic boundary is expressed as

$$\rho_c = \frac{a\sqrt{\sqrt{3}}}{\sqrt{2}\sqrt{\sin(\pi/3 - \psi_c)}} \exp\left(\frac{\sqrt{3}}{2}\psi_c\right). \tag{2.120}$$

Solving Eq. (2.118) for A and ρ_c results in

$$\rho_c^2 = \frac{\sqrt{3}\,[q - \sin(\psi_c + \pi/6)]}{\sin(\psi_c - \pi/3)}, \quad A = \sin\left(\psi_c + \frac{\pi}{6}\right) - q. \tag{2.121}$$

Eliminating ρ_c between Eqs. (2.120) and (2.121) yields the following relation that connects q and ψ_c

$$q = \sin\left(\psi_c + \frac{\pi}{6}\right) - \frac{a^2}{2}\exp\left(\sqrt{3}\psi_c\right). \tag{2.122}$$

The entire disc becomes plastic when $\rho_c = 1$. The equation for the corresponding value of ψ_c denoted by ψ_q follows from Eq. (2.120) in the form

$$\frac{a}{\sqrt{\sin(\pi/3 - \psi_q)}}\exp\left(\frac{\sqrt{3}}{2}\psi_q\right) = 1. \tag{2.123}$$

The corresponding value of q, q_q, is readily determined from Eq. (2.122). At any given value of q of the range, $q_e \le q \le q_q$, the solution to Eq. (2.122) supplies the corresponding value of ψ_c and, then, Eq. (2.120) the value of ρ_c. The distributions of the stresses in the elastic region, $\rho_c \le \rho \le 1$, follow from Eq. (1.29) in which A and B should be eliminated by means of Eqs. (2.115) and (2.121). The distributions of the stresses in the plastic region, $a \le \rho \le \rho_c$, are given by Eqs. (1.32) and (2.119) in parametric form with ψ being the parameter. It is convenient to put $p \equiv \psi_c$ in Eq. (1.16). It is seen from Eq. (2.119) that ψ is independent of ψ_c. Therefore, $\xi_r^e = \xi_\theta^e = \xi_z^e = 0$ and

$$\xi_r = \xi_r^p, \quad \xi_\theta = \xi_\theta^p, \quad \xi_z = \xi_z^p \tag{2.124}$$

in the plastic zone. In this case Eq. (1.70) transforms to

$$\frac{\partial \xi_\theta^p}{\partial \psi} + \sqrt{3}\xi_\theta^p = 0. \tag{2.125}$$

It has been taken into account here that $d\psi_0/dp = 0$. Equation (1.63) becomes

$$\frac{\xi_c}{k} = -\frac{dA}{d\psi_c}\frac{(1+\nu)}{\rho_c^2} + \frac{dB}{d\psi_c}(1-\nu). \tag{2.126}$$

It follows from Eqs. (2.115), (2.121) and (2.122) that

$$A = \frac{a^2}{2}\exp\left(\sqrt{3}\psi_c\right), \quad B = -\sin\left(\psi_c + \frac{\pi}{6}\right) \tag{2.127}$$

and, therefore,

$$\frac{dB}{d\psi_c} = -\cos\left(\psi_c + \frac{\pi}{6}\right), \quad \frac{dA}{d\psi_c} = \frac{\sqrt{3}a^2}{2}\exp\left(\sqrt{3}\psi_c\right). \tag{2.128}$$

Substituting Eqs. (2.128) and (2.120) into Eq. (2.126) gives

$$\frac{\xi_c}{k} = 2\sin\left(\psi_c - \frac{\pi}{3}\right). \tag{2.129}$$

Using Eq. (2.124) it is possible to represent the boundary condition (1.62) as $\xi_\theta^p = \xi_c$ for $\psi = \psi_c$. The solution to Eq. (2.125) satisfying this boundary condition is

$$\frac{\xi_\theta^p}{k} = 2\sin\left(\psi_c - \frac{\pi}{3}\right)\exp\left[-\sqrt{3}\left(\psi - \psi_c\right)\right]. \tag{2.130}$$

Here Eq. (2.129) has been used to eliminate ξ_c. Combining Eqs. (1.68) and (2.130) leads to

$$\frac{\xi_r^p}{k} = \frac{2\sin\left(\psi_c - \pi/3\right)\sin\left(\psi - \pi/6\right)}{\cos\psi}\exp\left[-\sqrt{3}\left(\psi - \psi_c\right)\right], \tag{2.131}$$
$$\frac{\xi_z^p}{k} = -\frac{2\sin\left(\psi_c - \pi/3\right)\sin\left(\psi + \pi/6\right)}{\cos\psi}\exp\left[-\sqrt{3}\left(\psi - \psi_c\right)\right].$$

The procedure for finding the strains is as follows. Let Υ be the value of ρ at which the strains should be calculated at $q = q_m$. Equation (2.122) gives

$$q_m = \sin\left(\psi_m + \frac{\pi}{6}\right) - \frac{a^2}{2}\exp\left(\sqrt{3}\psi_m\right). \tag{2.132}$$

Here ψ_m is the value of ψ_c at $q = q_m$. Equation (2.132) should be solved for ψ_m numerically. Then, the value of ρ_c at $q = q_m$ denoted by ρ_m is determined from Eq. (2.120) as

$$\rho_m = \frac{a\sqrt{\sqrt{3}}}{\sqrt{2}\sqrt{\sin\left(\pi/3 - \psi_m\right)}}\exp\left(\frac{\sqrt{3}}{2}\psi_m\right). \tag{2.133}$$

In the elastic region, $\rho_m \le \Upsilon \le 1$, the distributions of the strains follow from Eq. (1.60) in the form

$$\frac{\varepsilon_r^e}{k} = \frac{\varepsilon_r}{k} = \frac{A_m\left(1+\nu\right)}{\Upsilon^2} + B_m\left(1-\nu\right), \tag{2.134}$$
$$\frac{\varepsilon_\theta^e}{k} = \frac{\varepsilon_\theta}{k} = -\frac{A_m\left(1+\nu\right)}{\Upsilon^2} + B_m\left(1-\nu\right), \quad \frac{\varepsilon_z^e}{k} = \frac{\varepsilon_z}{k} = -2\nu B_m.$$

Having found ψ_m from Eq. (2.132) the values of A_m and B_m are determined by means of Eq. (2.127) as

$$A_m = \frac{a^2}{2} \exp\left(\sqrt{3}\psi_m\right), \quad B_m = -\sin\left(\psi_m + \frac{\pi}{6}\right). \tag{2.135}$$

In order to calculate the strains in the plastic region, $a \le \Upsilon \le \rho_m$, it is convenient to introduce the value of ψ_c at $\rho_c = \Upsilon$. This value of ψ_c is denoted by ψ_Υ. The equation for ψ_Υ follows from Eq. (2.120) in the form

$$\Upsilon = \frac{a\sqrt{\sqrt{3}}}{\sqrt{2}\sqrt{\sin\left(\pi/3 - \psi_\Upsilon\right)}} \exp\left(\frac{\sqrt{3}}{2}\psi_\Upsilon\right). \tag{2.136}$$

This equation should be solved numerically. Then, the corresponding values of $A = A_\Upsilon$ and $B = B_\Upsilon$ are found from Eq. (2.127) as

$$A_\Upsilon = \frac{a^2}{2} \exp\left(\sqrt{3}\psi_\Upsilon\right), \quad B_\Upsilon = -\sin\left(\psi_\Upsilon + \frac{\pi}{6}\right). \tag{2.137}$$

The elastic portions of the strains in the plastic zone are determined from Eq. (1.60) in the form

$$\begin{aligned} \frac{\varepsilon_r^e}{k} &= \frac{A_\Upsilon\left(1+\nu\right)}{\Upsilon^2} + B_\Upsilon\left(1-\nu\right), \\ \frac{\varepsilon_\theta^e}{k} &= -\frac{A_\Upsilon\left(1+\nu\right)}{\Upsilon^2} + B_\Upsilon\left(1-\nu\right), \quad \frac{\varepsilon_z^e}{k} = -2\nu B_\Upsilon. \end{aligned} \tag{2.138}$$

The plastic portions are given by

$$\varepsilon_r^p = \int_{\psi_\Upsilon}^{\psi_m} \xi_r^p d\psi_c, \quad \varepsilon_\theta^p = \int_{\psi_\Upsilon}^{\psi_m} \xi_\theta^p d\psi_c, \quad \varepsilon_z^p = \int_{\psi_\Upsilon}^{\psi_m} \xi_z^p d\psi_c. \tag{2.139}$$

Substituting Eqs. (2.130) and (2.131) into Eq. (2.139) and integrating yield

$$\begin{aligned} \frac{\varepsilon_r^p}{k} &= \sin\left(\psi_\Upsilon - \pi/6\right) - \frac{\sin\left(\psi_\Upsilon - \pi/6\right)\cos\psi_m}{\cos\psi_\Upsilon} \exp\left[\sqrt{3}\left(\psi_m - \psi_\Upsilon\right)\right], \\ \frac{\varepsilon_\theta^p}{k} &= \cos\psi_\Upsilon - \cos\psi_m \exp\left[\sqrt{3}\left(\psi_m - \psi_\Upsilon\right)\right], \\ \frac{\varepsilon_r^p}{k} &= \frac{\sin\left(\psi_\Upsilon + \pi/6\right)\cos\psi_m}{\cos\psi_\Upsilon} \exp\left[\sqrt{3}\left(\psi_m - \psi_\Upsilon\right)\right] - \sin\left(\psi_\Upsilon + \pi/6\right). \end{aligned} \tag{2.140}$$

It has been taken into account here that ψ is independent of ψ_c. Using the solution to Eqs. (2.136) and (2.137) the total strains in the plastic zone are determined from Eqs. (1.3), (2.138) and (2.140) at any given value of Υ.

2.2.2 Yield Criterion (1.8)

Substituting Eq. (2.113) into Eq. (1.38) yields

$$b_1 = \frac{q^2}{\left(1-a^2\right)^2}\left[1+\eta_1\left(\eta_1-\eta\right)\right]-\eta_1^2, \quad b_2 = \frac{2a^2q^2}{\left(1-a^2\right)^2}\left(1-\eta_1^2\right), \tag{2.141}$$

$$b_3 = \frac{a^4q^2}{\left(1-a^2\right)^2}\left[1+\eta_1\left(\eta_1+\eta\right)\right].$$

It is convenient to consider the cases $\eta_1 \leq 1$ and $\eta_1 > 1$ separately. Firstly, it is assumed that $\eta_1 \leq 1$. In this case $b_2 \geq 0$ and the plastic zone starts to develop from the inner radius of the disc (see Sect. 1.3.3). Then, Eq. (1.37) at $\rho = a$ and Eq. (2.141) combine to give

$$q_e = \frac{\eta_1\left(1-a^2\right)}{2}. \tag{2.142}$$

The material is elastic in the range $\rho_c \leq \rho \leq 1$. Therefore, Eq. (1.29) are valid in this range. However, A and B are not determined by Eq. (2.113). Nevertheless, the radial stress from Eq. (1.29) must satisfy the boundary condition (2.110). Then,

$$A + B = -q. \tag{2.143}$$

Equation (1.47) are satisfied in the range $a \leq \rho \leq \rho_c$. The boundary condition (2.111) and Eq. (1.47) for the radial stress combine to give

$$\sin\psi_a = 0 \tag{2.144}$$

where ψ_a is the value of ψ at $\rho = a$. Then, it follows from Eq. (1.47) that $p_\theta = -\sigma_0\cos\psi_a$ at $\rho = a$. It is evident that $\sigma_\theta < 0$ and, therefore, $p_\theta < 0$ at $\rho = a$. Hence $\cos\psi_a > 0$ and the solution of Eq. (2.144) is

$$\psi_a = 0. \tag{2.145}$$

It follows from Eqs. (1.9), (1.27), (1.47) and (2.143) that

$$q + A\left(1-\frac{1}{\rho_c^2}\right) = \frac{2\sin\psi_c}{\sqrt{4-\eta^2}}, \tag{2.146}$$

$$q + A\left(1+\frac{1}{\rho_c^2}\right) = \eta_1\left(\frac{\eta}{\sqrt{4-\eta^2}}\sin\psi_c+\cos\psi_c\right).$$

It is convenient to put $\psi_0 = \psi_a = 0$ and $\rho_0 = a$ in Eq. (1.50). Then,

$$\ln\left(\frac{\rho}{a}\right) = \frac{\eta_1\sqrt{4-\eta^2}}{2\left[1+\eta_1\left(\eta_1-\eta\right)\right]}\psi + \tag{2.147}$$
$$+ \frac{(2-\eta\eta_1)}{2\left[1+\eta_1\left(\eta_1-\eta\right)\right]}\ln\left[\frac{\eta_1\sqrt{4-\eta^2}}{\eta_1\sqrt{4-\eta^2}\cos\psi-(2-\eta\eta_1)\sin\psi}\right].$$

It follows from this equation and the definition for ψ_c that

$$\ln\left(\frac{\rho_c}{a}\right) = \frac{\eta_1\sqrt{4-\eta^2}}{2\left[1+\eta_1\left(\eta_1-\eta\right)\right]}\psi_c + \tag{2.148}$$
$$+ \frac{(2-\eta\eta_1)}{2\left[1+\eta_1\left(\eta_1-\eta\right)\right]}\ln\left[\frac{\eta_1\sqrt{4-\eta^2}}{\eta_1\sqrt{4-\eta^2}\cos\psi_c-(2-\eta\eta_1)\sin\psi_c}\right].$$

Solving Eq. (2.146) for ρ_c and A gives

$$\rho_c^2 = \frac{(2+\eta\eta_1)\sin\psi_c + \eta_1\sqrt{4-\eta^2}\cos\psi_c - 2q\sqrt{4-\eta^2}}{\eta_1\left(\eta\sin\psi_c+\sqrt{4-\eta^2}\cos\psi_c\right)-2\sin\psi_c}, \tag{2.149}$$
$$A = \frac{\sin\psi_c}{\sqrt{4-\eta^2}}\left(1+\frac{\eta\eta_1}{2}\right) + \frac{\eta_1}{2}\cos\psi_c - q.$$

The entire disc becomes plastic when $\rho_c = 1$. The corresponding value of ψ_c is denoted by ψ_q. It follows from Eq. (2.148) that the equation for ψ_q is

$$-\ln a = \frac{\eta_1\sqrt{4-\eta^2}}{2\left[1+\eta_1\left(\eta_1-\eta\right)\right]}\psi_q + \tag{2.150}$$
$$+ \frac{(2-\eta\eta_1)}{2\left[1+\eta_1\left(\eta_1-\eta\right)\right]}\ln\left[\frac{\eta_1\sqrt{4-\eta^2}}{\eta_1\sqrt{4-\eta^2}\cos\psi_q-(2-\eta\eta_1)\sin\psi_q}\right].$$

The corresponding value of q is determined from Eq. (2.146) as

$$q_q = \frac{2\sin\psi_q}{\sqrt{4-\eta^2}}. \tag{2.151}$$

For any given value of q of the range $q_e < q < q_q$, the corresponding values of ψ_c and ρ_c are found from the solution to Eqs. (2.148) and $(2.149)^1$. Then, A and B are determined from Eqs. (2.143) and $(2.149)^2$. The distributions of the radial and circumferential stresses are immediately found from Eq. (1.29) in the range $\rho_c \le \rho \le 1$ and from Eqs. (1.9), (1.47) and (2.147) in the range $a \le \rho \le \rho_c$. The latter is in parametric form with ψ being the parameter varying in the range $0 \le \psi \le \psi_c$.

It is convenient to put $p \equiv \psi_c$ in Eq. (1.16). It is seen from Eq. (2.147) that ψ is independent of ψ_c. Therefore, $\xi_r^e = \xi_\theta^e = \xi_z^e = 0$ and

$$\xi_r = \xi_r^p, \quad \xi_\theta = \xi_\theta^p, \quad \xi_z = \xi_z^p \tag{2.152}$$

in the plastic zone. In this case Eq. (1.80) transforms to

$$\frac{\partial \xi_\theta^p}{\partial \psi} = \frac{\left[\eta_1\left(\sqrt{4-\eta^2}\tan\psi - \eta\right) - 2\right]}{\left[\eta_1\sqrt{4-\eta^2} - (2-\eta\eta_1)\tan\psi\right]}\xi_\theta^p. \tag{2.153}$$

It has been taken into account here that $d\psi_0/dp = 0$ and $W_0(\psi)$ has been eliminated by means of Eq. (1.81). Equation (1.63) becomes

$$\frac{\xi_c}{k} = -\frac{dA}{d\psi_c}\frac{(1+\nu)}{\rho_c^2} + \frac{dB}{d\psi_c}(1-\nu). \tag{2.154}$$

Therefore, it is necessary to find the derivatives $dA/d\psi_c$ and $dB/d\psi_c$. It follows from Eqs. (2.143) and (2.149)[2] that

$$\frac{dB}{d\psi_c} = -\frac{dA}{d\psi_c} - \frac{dq}{d\psi_c}, \quad \frac{dA}{d\psi_c} = \frac{\cos\psi_c}{\sqrt{4-\eta^2}}\left(1+\frac{\eta\eta_1}{2}\right) - \frac{\eta_1}{2}\sin\psi_c - \frac{dq}{d\psi_c}. \tag{2.155}$$

The derivative $dq/d\psi_c$ is determined from Eq. (2.149)[1] in the form

$$\begin{aligned} 2\sqrt{4-\eta^2}\frac{dq}{d\psi_c} &= (2+\eta\eta_1)\cos\psi_c - \eta_1\sqrt{4-\eta^2}\sin\psi_c - \\ &- \left[(\eta\eta_1 - 2)\sin\psi_c + \eta_1\sqrt{4-\eta^2}\cos\psi_c\right]\frac{d\rho_c^2}{d\psi_c} - \\ &- \left[(\eta\eta_1 - 2)\cos\psi_c - \eta_1\sqrt{4-\eta^2}\sin\psi_c\right]\rho_c^2. \end{aligned} \tag{2.156}$$

The derivative $d\rho_c^2/d\psi_c$ involved in Eq. (2.156) is found from Eq. (2.148) as

$$\begin{aligned} \frac{d\rho_c^2}{\rho_c^2 d\psi_c} &= \frac{\eta_1\sqrt{4-\eta^2}}{1+\eta_1(\eta_1-\eta)} + \\ &+ \frac{(2-\eta\eta_1)}{[1+\eta_1(\eta_1-\eta)]}\frac{\left[\eta_1\sqrt{4-\eta^2}\sin\psi_c + (2-\eta\eta_1)\cos\psi_c\right]}{\left[\eta_1\sqrt{4-\eta^2}\cos\psi_c - (2-\eta\eta_1)\sin\psi_c\right]}. \end{aligned} \tag{2.157}$$

Substituting Eq. (2.157) into Eq. (2.156) and the resulting expression into Eq. (2.155)[2] yields

$$\frac{dA}{d\psi_c} = \frac{\rho_c^2}{2}\left[\frac{(\eta\eta_1 + 2)}{\sqrt{4-\eta^2}}\cos\psi_c - \eta_1 \sin\psi_c\right]. \tag{2.158}$$

Then, the derivative $dB/d\psi_c$ is found from Eqs. (2.155) as

$$\frac{dB}{d\psi_c} = \frac{\eta_1}{2}\sin\psi_c - \frac{\cos\psi_c}{\sqrt{4-\eta^2}}\left(1 + \frac{\eta\eta_1}{2}\right). \tag{2.159}$$

Substituting Eqs. (2.158) and (2.159) into Eq. (2.154) gives

$$\frac{\xi_c}{k} = \eta_1 \sin\psi_c - \frac{(2+\eta\eta_1)\cos\psi_c}{\sqrt{4-\eta^2}}. \tag{2.160}$$

Using Eq. (2.152) it is possible to represent the boundary condition (1.62) as $\xi_\theta^p = \xi_c$ for $\psi = \psi_c$. The solution of Eq. (2.153) satisfying this boundary condition can be written in terms of elementary functions. However, the final expression is cumbersome. Therefore, it is more convenient to represent the solution to this equation as

$$\xi_\theta^p = \xi_c \exp\int\limits_{\psi_c}^{\psi} \frac{\left[\eta_1\left(\sqrt{4-\eta^2}\tan\mu - \eta\right) - 2\right]}{\left[\eta_1\sqrt{4-\eta^2} - (2-\eta\eta_1)\tan\mu\right]} d\mu. \tag{2.161}$$

Here ξ_c should be eliminated by means of Eq. (2.160).

The procedure for finding the strains is as follows. Let Υ be the value of ρ at which the strains should be calculated at $q = q_m$. Eliminating ρ_c in Eq. (2.148) by means of Eq. (2.149) and putting $q = q_m$ leads to

$$\begin{aligned} &\ln\left[\frac{(2+\eta\eta_1)\sin\psi_m + \eta_1\sqrt{4-\eta^2}\cos\psi_m - 2q_m\sqrt{4-\eta^2}}{\eta_1\left(\eta\sin\psi_m + \sqrt{4-\eta^2}\cos\psi_m\right) - 2\sin\psi_m}\right] - 2\ln a = \\ &= \frac{\eta_1\sqrt{4-\eta^2}}{[1+\eta_1(\eta_1-\eta)]}\psi_m + \\ &+ \frac{(2-\eta\eta_1)}{[1+\eta_1(\eta_1-\eta)]}\ln\left[\frac{\eta_1\sqrt{4-\eta^2}}{\eta_1\sqrt{4-\eta^2}\cos\psi_m - (2-\eta\eta_1)\sin\psi_m}\right]. \end{aligned} \tag{2.162}$$

Here ψ_m is the value of ψ_c at $q = q_m$. Equation (2.162) should be solved for ψ_m numerically. Then, the value of ρ_c at $q = q_m$ denoted by ρ_m is determined from Eq. (2.149) as

$$\rho_m^2 = \frac{(2+\eta\eta_1)\sin\psi_m + \eta_1\sqrt{4-\eta^2}\cos\psi_m - 2q_m\sqrt{4-\eta^2}}{\eta_1\left(\eta\sin\psi_m + \sqrt{4-\eta^2}\cos\psi_m\right) - 2\sin\psi_m}. \tag{2.163}$$

In the elastic region, $\rho_m \le \Upsilon \le 1$, the distributions of the strains follow from Eq. (1.60) in the form

$$\begin{aligned} \frac{\varepsilon_r^e}{k} &= \frac{\varepsilon_r}{k} = \frac{A_m(1+\nu)}{\Upsilon^2} + B_m(1-\nu), \\ \frac{\varepsilon_\theta^e}{k} &= \frac{\varepsilon_\theta}{k} = -\frac{A_m(1+\nu)}{\Upsilon^2} + B_m(1-\nu) \quad \frac{\varepsilon_z^e}{k} = \frac{\varepsilon_z}{k} = -2\nu B_m. \end{aligned} \tag{2.164}$$

Having found ψ_m from Eq. (2.162) the values of A_m and B_m are determined by means of Eqs. (2.143) and (2.149) as

$$\begin{aligned} A_m &= \frac{\sin\psi_m}{\sqrt{4-\eta^2}}\left(1+\frac{\eta\eta_1}{2}\right) + \frac{\eta_1}{2}\cos\psi_m - q_m, \\ B_m &= -\frac{\sin\psi_m}{\sqrt{4-\eta^2}}\left(1+\frac{\eta\eta_1}{2}\right) - \frac{\eta_1}{2}\cos\psi_m. \end{aligned} \tag{2.165}$$

In order to calculate the strains in the plastic region, $a \le \Upsilon \le \rho_m$, it is convenient to introduce the value of ψ_c at $\rho_c = \Upsilon$. This value of ψ_c is denoted by ψ_Υ. The equation for ψ_Υ follows from Eq. (2.148) in the form

$$\begin{aligned} 2\ln\left(\frac{\Upsilon}{a}\right) &= \frac{\eta_1\sqrt{4-\eta^2}}{[1+\eta_1(\eta_1-\eta)]}\psi_\Upsilon \\ &+ \frac{(2-\eta\eta_1)}{[1+\eta_1(\eta_1-\eta)]}\ln\left[\frac{\eta_1\sqrt{4-\eta^2}}{\eta_1\sqrt{4-\eta^2}\cos\psi_\Upsilon - (2-\eta\eta_1)\sin\psi_\Upsilon}\right]. \end{aligned} \tag{2.166}$$

This equation should be solved numerically. Then, the corresponding values of $q = q_\Upsilon$, $A = A_\Upsilon$ and $B = B_\Upsilon$ are found from Eqs. (2.143) and (2.149) as

$$\begin{aligned} 2\sqrt{4-\eta^2}q_\Upsilon &= \Upsilon^2\left[2\sin\psi_\Upsilon - \eta_1\left(\eta\sin\psi_\Upsilon + \sqrt{4-\eta^2}\cos\psi_\Upsilon\right)\right] \\ &+ (2+\eta\eta_1)\sin\psi_\Upsilon + \eta_1\sqrt{4-\eta^2}\cos\psi_\Upsilon, \\ A_\Upsilon &= \frac{\sin\psi_\Upsilon}{\sqrt{4-\eta^2}}\left(1+\frac{\eta\eta_1}{2}\right) + \frac{\eta_1}{2}\cos\psi_\Upsilon - q_\Upsilon, \\ B_\Upsilon &= -\frac{\sin\psi_\Upsilon}{\sqrt{4-\eta^2}}\left(1+\frac{\eta\eta_1}{2}\right) - \frac{\eta_1}{2}\cos\psi_\Upsilon. \end{aligned} \tag{2.167}$$

The elastic portions of the strains in the plastic zone are determined from Eq. (1.60) in the form

$$\frac{\varepsilon_r^e}{k} = \frac{A_\Upsilon (1+\nu)}{\Upsilon^2} + B_\Upsilon (1-\nu), \tag{2.168}$$
$$\frac{\varepsilon_\theta^e}{k} = -\frac{A_\Upsilon (1+\nu)}{\Upsilon^2} + B_\Upsilon (1-\nu), \quad \frac{\varepsilon_z^e}{k} = -2\nu B_\Upsilon .$$

The plastic portions are given by

$$\varepsilon_r^p = \int_{\psi_\Upsilon}^{\psi_m} \xi_r^p d\psi_c, \quad \varepsilon_\theta^p = \int_{\psi_\Upsilon}^{\psi_m} \xi_\theta^p d\psi_c, \quad \varepsilon_z^p = \int_{\psi_\Upsilon}^{\psi_m} \xi_z^p d\psi_c. \tag{2.169}$$

Substituting Eqs. (2.160) and (2.161) into Eq. (2.169) for ε_θ^p yields

$$\varepsilon_\theta^p = \int_{\psi_\Upsilon}^{\psi_m} \begin{array}{l} \left[\eta_1 \sin\mu_1 - \frac{(2+\eta\eta_1)}{\sqrt{4-\eta^2}} \cos\mu_1\right] \times \\ \left\{\exp \int\limits_{\mu_1}^{\psi_\Upsilon} \frac{\left[\eta_1\left(\sqrt{4-\eta^2}\tan\mu - \eta\right) - 2\right]}{\left[\eta_1\sqrt{4-\eta^2} - (2-\eta\eta_1)\tan\mu\right]} d\mu\right\} d\mu_1 \end{array}. \tag{2.170}$$

Eliminating ξ_r^p and ξ_z^p in Eq. (2.169) by means of Eq. (1.78), taking into account that ψ is independent of ψ_c and integrating lead to

$$\varepsilon_r^p = \frac{\eta_1}{2}\left(\sqrt{4-\eta^2}\tan\psi_\Upsilon - \eta\right)\varepsilon_\theta^p, \tag{2.171}$$
$$\varepsilon_z^p = -\frac{\left(2 - \eta\eta_1 + \eta_1\sqrt{4-\eta^2}\tan\psi_\Upsilon\right)}{2}\varepsilon_\theta^p .$$

Here the strain ε_θ^p can be eliminated by means of Eq. (2.170). Using the solution to Eqs. (2.166) and (2.167) the total strains in the plastic zone are determined from Eqs. (1.3), (2.168), (2.170) and (2.171) at any given value of Υ.

It is now assumed that $\eta_1 > 1$. The case corresponding to Eq. (1.44) is treated in the same manner as the case $\eta_1 \le 1$ since the plastic zone starts to develop from the inner radius of the disc. However, another plastic zone can start to develop from the outer radius of the disc if q is large enough. The corresponding condition follows from Eqs. (1.36) and (2.143) in the form

$$4A^2 + 2(2-\eta\eta_1) Aq + [1 + \eta_1(\eta_1 - \eta)] q^2 \le \eta_1^2. \tag{2.172}$$

Since the solution to Eqs. (2.148) and (2.149) supplies A and q as functions of ψ_c, Eq. (2.172) imposes a restriction on the value of ψ_c. The corresponding restrictions on the values of ρ_c and q follow from Eqs. (2.148) and (2.149). If the inequality

(2.172) is not satisfied then it is necessary to find a solution with two plastic zones. This solution is beyond the scope of the present monograph.

Assume that Eq. (1.45) is satisfied. Then, the plastic zone starts to develop from the outer radius of the disc. Equation (1.37) at $\rho = 1$, (1.38) and (2.113) combine to give

$$q_e = \frac{\eta_1 \left(1 - a^2\right)}{\sqrt{1 - \eta\eta_1 + \eta_1^2 + 2\left(1 - \eta_1^2\right) a^2 + \left(1 + \eta\eta_1 + \eta_1^2\right) a^4}}. \tag{2.173}$$

Since the plastic zone starts to develop from the outer radius of the disc, the material is elastic in the range $a \le \rho \le \rho_c$. Therefore, Eq. (1.29) are valid in this range. However, A and B are not determined by Eq. (2.113). Nevertheless, the radial stress from Eq. (1.29) must satisfy the boundary condition (2.111). Then,

$$B = -\frac{A}{a^2}. \tag{2.174}$$

The radial stress determined by Eq. (1.47) must satisfy the boundary condition (2.110). Therefore,

$$2 \sin \psi_b = q\sqrt{4 - \eta^2}. \tag{2.175}$$

Here ψ_b is the value of ψ at $\rho = 1$. It is evident that $\sigma_\theta < 0$ at $\rho = 1$. Therefore, it is seen from Eqs. (1.9) and (1.47) that

$$\frac{\eta \sin \psi_b}{\sqrt{4 - \eta^2}} + \cos \psi_b > 0.$$

This inequality and Eq. (2.175) allow a unique value of ψ_b to be determined. It is convenient to put $\psi_0 = \psi_b = p$ and $\rho_0 = 1$ in Eqs. (1.16) and (1.50). Then,

$$\ln \rho = \frac{\eta_1\sqrt{4 - \eta^2}}{2\left[1 + \eta_1 \left(\eta_1 - \eta\right)\right]} \left(\psi - \psi_b\right) + \frac{\left(2 - \eta\eta_1\right)}{2\left[1 + \eta_1 \left(\eta_1 - \eta\right)\right]} \ln \left[\frac{\eta_1\sqrt{4 - \eta^2} \cos \psi_b - \left(2 - \eta\eta_1\right) \sin \psi_b}{\eta_1\sqrt{4 - \eta^2} \cos \psi - \left(2 - \eta\eta_1\right) \sin \psi}\right]. \tag{2.176}$$

It follows from this equation and the definition for ψ_c that

$$\ln \rho_c = \frac{\eta_1\sqrt{4 - \eta^2}}{2\left[1 + \eta_1 \left(\eta_1 - \eta\right)\right]} \left(\psi_c - \psi_b\right) + \frac{\left(2 - \eta\eta_1\right)}{2\left[1 + \eta_1 \left(\eta_1 - \eta\right)\right]} \ln \left[\frac{\eta_1\sqrt{4 - \eta^2} \cos \psi_b - \left(2 - \eta\eta_1\right) \sin \psi_b}{\eta_1\sqrt{4 - \eta^2} \cos \psi_c - \left(2 - \eta\eta_1\right) \sin \psi_c}\right]. \tag{2.177}$$

It follows from Eqs. (1.9), (1.27), (1.29), (1.47), and (2.174) that

$$\left(\frac{1}{\rho_c^2} - \frac{1}{a^2}\right) A = -\frac{2 \sin \psi_c}{\sqrt{4 - \eta^2}}, \tag{2.178}$$

$$\left(\frac{1}{\rho_c^2} + \frac{1}{a^2}\right) A = \eta_1 \left(\frac{\eta \sin \psi_c}{\sqrt{4 - \eta^2}} + \cos \psi_c\right).$$

Solving these equations for A and ρ_c results in

$$A = \frac{a^2}{2}\left[\frac{(\eta\eta_1 + 2) \sin \psi_c}{\sqrt{4 - \eta^2}} + \eta_1 \cos \psi_c\right], \tag{2.179}$$

$$\rho_c^2 = -a^2 \frac{\left(2 + \eta\eta_1 + \eta_1 \sqrt{4 - \eta^2} \cot \psi_c\right)}{\left(2 - \eta\eta_1 - \eta_1 \sqrt{4 - \eta^2} \cot \psi_c\right)}.$$

Substituting Eq. (2.179)[2] into Eq. (2.177) gives the following equation

$$2 \ln a + \ln\left(\frac{\eta_1 \sqrt{4 - \eta^2} \cot \psi_c + \eta\eta_1 + 2}{\eta_1 \sqrt{4 - \eta^2} \cot \psi_c + \eta\eta_1 - 2}\right) = \frac{\eta_1 \sqrt{4 - \eta^2}}{[1 + \eta_1 (\eta_1 - \eta)]} (\psi_c - \psi_b) +$$

$$+ \frac{(2 - \eta\eta_1)}{[1 + \eta_1 (\eta_1 - \eta)]} \ln\left[\frac{\eta_1 \sqrt{4 - \eta^2} \cos \psi_b - (2 - \eta\eta_1) \sin \psi_b}{\eta_1 \sqrt{4 - \eta^2} \cos \psi_c - (2 - \eta\eta_1) \sin \psi_c}\right]. \tag{2.180}$$

The solution to this equation supplies ψ_c as a function of ψ_b. The entire disc becomes plastic when $\rho_c = a$. The equation for the corresponding value of ψ_c denoted by ψ_q follows from Eq. (2.178) in the form $\sin \psi_q = 0$. Since $\sigma_\theta < 0$ at $\rho = a$, the solution to this equation is $\psi_q = 0$, as follows from Eqs. (1.9) and (1.47). The value of q at $\psi_c = \psi_q$ is found from the solution to Eqs. (2.180) and (2.172). This value of q is denoted by q_q. At any value of q in the range $q_e \le q \le q_q$, the corresponding values of ψ_b and ψ_c are determined from Eqs. (2.172) and (2.180). Then, the values of ρ_c, A and B immediately follow from Eqs. (2.171), and (2.179). The distributions of stresses are found from Eqs. (1.29) in the range $a \le \rho \le \rho_c$ and from Eqs. (1.9), (1.47) and (2.176) in the range $\rho_c \le \rho \le 1$. The latter is in parametric form with ψ being the parameter.

Replacing ψ_0 with ψ_b in Eq. (1.83) and using Eq. (1.79) lead to

$$\frac{d\psi}{\eta_1 \sqrt{4 - \eta^2} - (2 - \eta\eta_1) \tan \psi} = \frac{d\psi_b}{\eta_1 \sqrt{4 - \eta^2} - (2 - \eta\eta_1) \tan \psi_b}.$$

Integrating this equation gives

$$\eta_1\sqrt{4-\eta^2}\,(\psi-\psi_b)+(2-\eta\eta_1)\ln\left[\frac{\eta_1\sqrt{4-\eta^2}\cos\psi_b-(2-\eta\eta_1)\sin\psi_b}{\eta_1\sqrt{4-\eta^2}\cos\psi-(2-\eta\eta_1)\sin\psi}\right]=$$
$$=2\left[1+\eta_1\left(\eta_1-\eta\right)\right]\ln C \tag{2.181}$$

where C is constant on each characteristic curve. Comparing Eqs. (2.176) and (2.181) shows that $C=\rho$. Therefore, integrating along the characteristics is equivalent to integrating at fixed values of ρ. Since $p=\psi_b$, it is possible to rewrite Eq. (1.84) as

$$\frac{d\varepsilon_r}{d\psi_b}=\xi_r,\quad \frac{d\varepsilon_\theta}{d\psi_b}=\xi_\theta,\quad \frac{d\varepsilon_z}{d\psi_b}=\xi_z. \tag{2.182}$$

Differentiating Eqs. (2.171) and (2.179) for A with respect to ψ_b gives

$$\frac{dA}{d\psi_b}=\frac{a^2}{2}\left[\frac{(\eta\eta_1+2)\cos\psi_c}{\sqrt{4-\eta^2}}-\eta_1\sin\psi_c\right]\frac{d\psi_c}{d\psi_b}, \tag{2.183}$$
$$\frac{dB}{d\psi_b}=-\frac{1}{2}\left[\frac{(\eta\eta_1+2)\cos\psi_c}{\sqrt{4-\eta^2}}-\eta_1\sin\psi_c\right]\frac{d\psi_c}{d\psi_b}.$$

The derivative $d\psi_c/d\psi_b$ is found from Eq. (2.182) as

$$\frac{d\psi_c}{d\psi_b}=\frac{V_2+V_3\left(\psi_c,\,\psi_b\right)}{V_0\left(\psi_c\right)+V_1\left(\psi_c,\,\psi_b\right)} \tag{2.184}$$

where

$$V_0\left(\psi_c\right)=\frac{4\eta_1\sqrt{4-\eta^2}}{\left[\left(\eta_1\sqrt{4-\eta^2}\cos\psi_c+\eta\eta_1\sin\psi_c\right)^2-4\sin^2\psi_c\right]}-\frac{\eta_1\sqrt{4-\eta^2}}{1+\eta_1\left(\eta_1-\eta\right)},$$

$$V_1\left(\psi_c,\,\psi_b\right)=\frac{(\eta\eta_1-2)}{\left[1+\eta_1\left(\eta_1-\eta\right)\right]}\times$$
$$\times\frac{\left[\begin{array}{c}\eta_1\left(2-\eta\eta_1\right)\sqrt{4-\eta^2}\cos\left(\psi_c+\psi_b\right)-\\ -(2-\eta\eta_1)^2\cos\psi_c\sin\psi_b+\left(4-\eta^2\right)\eta_1^2\cos\psi_b\sin\psi_c\end{array}\right]}{\left[\sqrt{4-\eta^2}\eta_1\cos\psi_b-\left(2-\eta\eta_1\right)\sin\psi_b\right]\left[\sqrt{4-\eta^2}\eta_1\cos\psi_c-\left(2-\eta\eta_1\right)\sin\psi_c\right]},$$

$$V_2=-\frac{\eta_1\sqrt{4-\eta^2}}{1+\eta_1\left(\eta_1-\eta\right)},\quad V_3\left(\psi_c,\,\psi_b\right)=\frac{\left(2-\eta\eta_1\right)}{\left[1+\eta_1\left(\eta_1-\eta\right)\right]}\times$$
$$\times\frac{\left[\begin{array}{c}\sqrt{4-\eta^2}\eta_1\left(\eta\eta_1-2\right)\cos\left(\psi_c+\psi_b\right)-\\ -\left(4-\eta^2\right)\eta_1^2\cos\psi_c\sin\psi_b+(2-\eta\eta_1)^2\cos\psi_b\sin\psi_c\end{array}\right]}{\left[\sqrt{4-\eta^2}\eta_1\cos\psi_b-\left(2-\eta\eta_1\right)\sin\psi_b\right]\left[\sqrt{4-\eta^2}\eta_1\cos\psi_c-\left(2-\eta\eta_1\right)\sin\psi_c\right]}.$$

Substituting Eq. (2.183) into Eq. (1.63) yields

$$\frac{\xi_c}{k} = -\frac{a^2}{2}\left[\frac{(1+\nu)}{\rho_c^2} + \frac{(1-\nu)}{a^2}\right]\left[\frac{(\eta\eta_1+2)\cos\psi_c}{\sqrt{4-\eta^2}} - \eta_1 \sin\psi_c\right]\frac{d\psi_c}{d\psi_b}. \quad (2.185)$$

Using Eqs. (2.184) and (2.185) the right hand side of Eq. (1.82) is represented as a function of ψ, ψ_b and ψ_c. Eliminating ψ_c by means of the solution to Eq. (2.180) supplies the right hand side of this equation as a function of ψ and ψ_b. Further eliminating ψ by means of the solution to Eq. (2.181) at any given value of $C = \rho$ determines the right hand side of Eq. (2.182) for ε_θ as a function of ψ_b. This function is denoted by $E_\theta(\psi_b)$. Therefore, Eq. (2.182) for ε_θ can be integrated numerically. In particular, the value of ε_θ at $\psi_b = \psi_m$ and $\rho = C$ is given by

$$\varepsilon_\theta = \int_{\psi_i}^{\psi_m} E_\theta(\psi_b)\, d\psi_b + E_\theta^e. \quad (2.186)$$

Here ψ_m is prescribed and Eq. (2.186) supplies ε_θ in the plastic zone. It is seen from Eq. (2.175) that prescribing the value of ψ_m is equivalent to prescribing the value of q. The procedure to find ψ_i and E_θ^e is as follows. The value of E_θ^e is the circumferential strain at $\rho = \rho_c = C$ and ψ_i is the value of ψ_b at $\rho_c = C$. It is seen from Eq. (1.26)[1] that E_θ^e is determined from the solution in the elastic zone. Alternatively, since the stresses are continuous across the elastic/plastic boundary, the value of E_θ^e can be found from Eqs. (1.1) and (1.47). Let ψ_C be the value of ψ_c at $\rho_c = C$. Then, the equation for ψ_C follows from Eq. (2.179) as

$$C^2 = -a^2\frac{\left(2+\eta\eta_1+\eta_1\sqrt{4-\eta^2}\cot\psi_C\right)}{\left(2-\eta\eta_1-\eta_1\sqrt{4-\eta^2}\cot\psi_C\right)}. \quad (2.187)$$

Having found the value of ψ_C the value of E_θ^e is determined from Eqs. (1.1) and (1.47) at $\psi = \psi_C$ as

$$\frac{E_\theta^e}{k} = \frac{(2\nu-\eta\eta_1)}{\sqrt{4-\eta^2}}\sin\psi_C - \eta_1\cos\psi_C. \quad (2.188)$$

Substituting $\psi_c = \psi_C$ into Eq. (2.180) and solving this equation for ψ_b supplies the value of ψ_i.

The distributions of ε_r and ε_z in the plastic zone are determined in a similar manner. In particular, using Eqs. (1.22) and (1.78)

$$\xi_r = \xi_r^e + \xi_r^p = \xi_r^e + \frac{\xi_\theta^p \eta_1}{2}\left(\sqrt{4-\eta^2}\tan\psi - \eta\right) =$$
$$= \xi_r^e + \frac{\left(\xi_\theta - \xi_\theta^e\right)\eta_1}{2}\left(\sqrt{4-\eta^2}\tan\psi - \eta\right), \tag{2.189}$$
$$\xi_z = \xi_z^e + \xi_z^p = \xi_z^e - \frac{\xi_\theta^p}{2}\left(\eta_1\sqrt{4-\eta^2}\tan\psi + 2 - \eta\eta_1\right) =$$
$$= \xi_r^e - \frac{\left(\xi_\theta - \xi_\theta^e\right)}{2}\left(\eta_1\sqrt{4-\eta^2}\tan\psi + 2 - \eta\eta_1\right).$$

Using Eqs. (1.76), (1.79) and the function $E_\theta\left(\psi_b\right)$ the right hand sides of Eq. (2.189) are expressed in terms of ψ and ψ_b. Eliminating ψ by means of the solution to Eq. (2.181) at a given value of C determines the right hand sides of these equations as functions of ψ_b. These functions are denoted by $E_r\left(\psi_b\right)$ and $E_z\left(\psi_b\right)$. Equation (2.182) for ε_r and ε_z can be integrated numerically. In particular,

$$\varepsilon_r = \int\limits_{\psi_i}^{\psi_m} E_r\left(\psi_b\right)d\psi_b + E_r^e, \quad \varepsilon_z = \int\limits_{\psi_i}^{\psi_m} E_z\left(\psi_b\right)d\psi_b + E_z^e. \tag{2.190}$$

These equations supply ε_r and ε_z in the plastic zone. In Eq. (2.190), E_r^e and E_z^e are the radial and axial strains, respectively, at $\rho = \rho_c = C$. These strains are determined from Eq. (1.61) at $\rho = \rho_c = C$, (2.174) and (2.179) at $\psi_c = \psi_C$ or from Eqs. (1.1) and (1.47) at $\psi = \psi_C$. As a result,

$$\frac{E_r^e}{k} = \frac{\left(\nu\eta\eta_1 - 2\right)}{\sqrt{4-\eta^2}}\sin\psi_C + \nu\eta_1\cos\psi_C, \tag{2.191}$$
$$\frac{E_z^e}{k} = \nu\left[\frac{\left(\eta\eta_1 + 2\right)\sin\psi_C}{\sqrt{4-\eta^2}} + \eta_1\cos\psi_C\right]$$

Having found the distribution of the total strains in the plastic zone it is possible to determine their plastic portion by means of Eq. (1.3) in which the elastic strains should be eliminated using Eqs. (1.1), (1.9), (1.47) and (2.176). The total strains in the elastic zone follow from Eq. (1.61) in which A and B should be eliminated by means of Eqs. (2.174) and (2.179). The value of ψ_c involved in Eq. (2.179) is determined from Eq. (2.180) assuming that $\psi_b = \psi_m$.

A restriction on the solution found is that another plastic zone can start to develop from the inner radius of the disc. Substituting Eq. (2.174) into Eq. (1.36) at $\rho = a$ and using Eq. (2.179) yield

$$\frac{\left(\eta\eta_1 + 2\right)\sin\psi_c}{\sqrt{4-\eta^2}} + \eta_1\left(\cos\psi_c - 1\right) = 0. \tag{2.192}$$

If this equation has such a solution for ψ_c that the corresponding value of ρ_c is in the range $a < \rho_c < 1$ then the domain of validity of the solution with one plastic zone has been found. The corresponding value of q is denoted by q_m. The value of q_m is determined from Eqs. (2.172) and (2.180). A solution with two plastic zones is required in the range $q > q_m$. A similar solution is required from the beginning of plastic yielding if Eq. (1.46) is satisfied. Such solutions are beyond the scope of the present monograph.

2.2.3 Yield Criterion (1.11)

Substituting Eq. (2.113) into Eq. (1.52) results in

$$q_e = \frac{3\left(1-a^2\right)}{2\left(3-\alpha\right)}. \tag{2.193}$$

Since the plastic zone starts to develop from the inner radius of the disc, the material is elastic in the range $\rho_c \leq \rho \leq 1$. Therefore, Eq. (1.29) are valid in this range. However, A and B are not determined by Eq. (2.113). Nevertheless, the radial stress determined by Eq. (1.29) must satisfy the boundary condition (2.110). Then,

$$A + B = -q. \tag{2.194}$$

The boundary condition (2.111) and the solution (1.53) for σ_r give

$$3\beta_0 - \frac{\beta_1}{2}\left(1 + 3\sqrt{3}\beta_1\right)\sin\psi_a + \frac{\sqrt{3}\beta_1}{2}\left(1 - \sqrt{3}\beta_1\right)\cos\psi_a = 0. \tag{2.195}$$

Here ψ_a is the value of ψ at $\rho = a$. It is evident that $\sigma_\theta < 0$ at $\rho = a$. Therefore, it follows from the solution (1.53) for σ_θ that

$$3\beta_0 + \frac{\beta_1}{2}\left(1 - 3\sqrt{3}\beta_1\right)\sin\psi_a - \frac{\sqrt{3}\beta_1}{2}\left(1 + \sqrt{3}\beta_1\right)\cos\psi_a < 0. \tag{2.196}$$

It is seen from Eqs. (2.195) and (2.196) that

$$\sin\left(\frac{\pi}{3} - \psi_a\right) > 0. \tag{2.197}$$

Equations (2.195) and (2.197) allow a unique value of ψ_a to be determined. It is seen from Eq. (2.195) that ψ_a is constant. It follows from Eqs. (1.27), (1.29), (1.53) and (2.194) that

$$3\beta_0 - \frac{\beta_1}{2}\left(1 + 3\sqrt{3}\beta_1\right)\sin\psi_c + \frac{\sqrt{3}\beta_1}{2}\left(1 - \sqrt{3}\beta_1\right)\cos\psi_c = A\left(\frac{1}{\rho_c^2} - 1\right) - q,$$
$$3\beta_0 + \frac{\beta_1}{2}\left(1 - 3\sqrt{3}\beta_1\right)\sin\psi_c - \frac{\sqrt{3}\beta_1}{2}\left(1 + \sqrt{3}\beta_1\right)\cos\psi_c = -A\left(\frac{1}{\rho_c^2} + 1\right) - q. \tag{2.198}$$

It is convenient to put $\psi_0 = \psi_a$ and $\rho_0 = a$ in Eq. (1.57). Then, this equation becomes

$$\rho = a\exp\left[\frac{3\beta_1}{2}(\psi - \psi_a)\right]\sqrt{\frac{\sin(\psi_a - \pi/3)}{\sin(\psi - \pi/3)}}. \tag{2.199}$$

Using this equation and the definition for ψ_c the radius of the elastic/plastic boundary is expressed as

$$\rho_c = a\exp\left[\frac{3\beta_1}{2}(\psi_c - \psi_a)\right]\sqrt{\frac{\sin(\psi_a - \pi/3)}{\sin(\psi_c - \pi/3)}}. \tag{2.200}$$

Solving Eq. (2.198) for A and ρ_c results in

$$\rho_c^2 = \frac{q + 3\beta_0 - 3\beta_1^2\sin(\psi_c + \pi/6)}{\beta_1\sin(\psi_c - \pi/3)}, \quad A = 3\beta_1^2\sin\left(\psi_c + \frac{\pi}{6}\right) - 3\beta_0 - q. \tag{2.201}$$

Eliminating ρ_c between Eqs. (2.200) and (2.201) yields the following relation that connects q and ψ_c

$$q = 3\beta_1^2\sin\left(\psi_c + \frac{\pi}{6}\right) + a^2\beta_1\exp\left[3\beta_1(\psi_c - \psi_a)\right]\sin\left(\psi_a - \frac{\pi}{3}\right) - 3\beta_0. \tag{2.202}$$

The entire disc becomes plastic when $\rho_c = 1$. The equation for the corresponding value of ψ_c denoted by ψ_q follows from Eq. (2.200) in the form

$$a\exp\left[\frac{3\beta_1}{2}(\psi_q - \psi_a)\right]\sqrt{\frac{\sin(\psi_a - \pi/3)}{\sin(\psi_q - \pi/3)}} = 1. \tag{2.203}$$

It is convenient to put $p \equiv \psi_c$ in Eq. (1.16). It is seen from Eqs. (2.195) and (2.199) that ψ is independent of ψ_c. Therefore, $\xi_r^e = \xi_\theta^e = \xi_z^e = 0$ and

$$\xi_r = \xi_r^p, \quad \xi_\theta = \xi_\theta^p, \quad \xi_z = \xi_z^p \tag{2.204}$$

in the plastic zone. In this case Eq. (1.88) transforms to

$$\frac{\partial\xi_\theta^p}{\partial\psi} = -\frac{\beta_1\left[\left(1 + 3\sqrt{3}\beta_1\right)\cos\psi + \sqrt{3}\left(1 - \sqrt{3}\beta_1\right)\sin\psi\right]}{\left[\beta_1\left(\sqrt{3} + \beta_1\right)\cos\psi + \beta_1\left(\sqrt{3}\beta_1 - 1\right)\sin\psi - 2\beta_0\right]}\xi_\theta^p. \tag{2.205}$$

It has been taken into account here that $d\psi_0/dp = 0$. Also, $W_0(\psi)$ has been eliminated by means of Eq. (1.89). Equation (2.205) is valid for plastically incompressible materials. The corresponding equation for plastically compressible materials follows from Eq. (1.94) in the form

$$\frac{\partial \xi_\theta^p}{\partial \psi} + 3\beta_1 \xi_\theta^p = 0. \tag{2.206}$$

Equation (1.63) becomes

$$\frac{\xi_c}{k} = -\frac{dA}{d\psi_c}\frac{(1+\nu)}{\rho_c^2} + \frac{dB}{d\psi_c}(1-\nu). \tag{2.207}$$

It follows from Eqs. (2.194), (2.201) and (2.202) that

$$A = \beta_1 a^2 \exp\left[3\beta_1\psi_c - \psi_a\right]\sin\left(\frac{\pi}{3} - \psi_a\right), \quad B = 3\left[\beta_0 - \beta_1^2 \sin\left(\psi_c + \frac{\pi}{6}\right)\right] \tag{2.208}$$

and, therefore,

$$\frac{dB}{d\psi_c} = -3\beta_1^2\cos\left(\psi_c + \frac{\pi}{6}\right), \quad \frac{dA}{d\psi_c} = 3\beta_1^2 a^2 \exp\left[3\beta_1\psi_c - \psi_a\right]\sin\left(\frac{\pi}{3} - \psi_a\right). \tag{2.209}$$

Substituting Eqs. (2.200) and (2.209) into Eq. (2.207) gives

$$\frac{\xi_c}{k} = 3\beta_1^2 \sin\left(\psi_c - \frac{\pi}{3}\right)\left[1 - \nu - \frac{\sqrt{3}(1+\nu)\exp(3\beta_1\psi_a)}{2\sin(\psi_a - \pi/3)}\right] \tag{2.210}$$

Using Eq. (2.204) it is possible to represent the boundary condition (1.62) as $\xi_\theta^p = \xi_c$ for $\psi = \psi_c$. The solution to Eq. (2.205) satisfying this boundary condition can be written as

$$\xi_\theta^p = \xi_c \exp\left[-\beta_1 \Lambda(\psi, \psi_c)\right], \tag{2.211}$$

$$\Lambda(\psi, \psi_c) = \int_{\psi_c}^{\psi} \frac{\left[\left(1 + 3\sqrt{3}\beta_1\right)\cos\mu + \sqrt{3}\left(1 - \sqrt{3}\beta_1\right)\sin\mu\right]}{\left[\beta_1\left(\sqrt{3} + \beta_1\right)\cos\mu + \beta_1\left(\sqrt{3}\beta_1 - 1\right)\sin\mu - 2\beta_0\right]} d\mu.$$

The integral in this solution can be evaluated in terms of elementary functions. However, the final expression is very cumbersome. Equation (2.211) is valid for plastically incompressible materials. The corresponding equation for plastically compressible materials follows from Eq. (2.206) in the form

$$\frac{\xi_\theta^p}{k} = \frac{\xi_c}{k}\exp\left[3\beta_1(\psi_c - \psi)\right]. \tag{2.212}$$

Combining Eqs. (1.86) and (2.211) leads to

$$\xi_r^p = \xi_c \exp\left[-\beta_1 \Lambda\left(\psi,\ \psi_c\right)\right]\left[\frac{\beta_1\left(\sqrt{3}-\beta_1\right)\cos\psi - \beta_1\left(1+\sqrt{3}\beta_1\right)\sin\psi + 2\beta_0}{\beta_1\left(1-\sqrt{3}\beta_1\right)\sin\psi - \beta_1\left(\sqrt{3}+\beta_1\right)\cos\psi + 2\beta_0}\right], \tag{2.213}$$

$$\xi_z^p = 2\xi_c \exp\left[-\beta_1 \Lambda\left(\psi,\ \psi_c\right)\right]\left[\frac{\beta_1^2\left(\cos\psi + \sqrt{3}\sin\psi\right) - 2\beta_0}{\beta_1\left(1-\sqrt{3}\beta_1\right)\sin\psi - \beta_1\left(\sqrt{3}+\beta_1\right)\cos\psi + 2\beta_0}\right]$$

in the case of plastically incompressible materials. The corresponding equations for plastically compressible materials follow from Eqs. (1.93) and (2.212) in the form

$$\xi_r^p = \xi_c \exp\left[3\beta_1\left(\psi_c - \psi\right)\right]\frac{\left[3\beta_1 \sin\left(\psi - \pi/3\right) + \sin\left(\psi + \pi/6\right)\right]}{\left[\sin\left(\psi + \pi/6\right) - 3\beta_1 \sin\left(\psi - \pi/3\right)\right]}, \tag{2.214}$$

$$\xi_z^p = -\frac{2\beta_1^2\xi_c}{3}\exp\left[3\beta_1\left(\psi_c - \psi\right)\right]\frac{\left[9\alpha + \left(9 + 2\alpha^2\right)\sin\left(\psi + \pi/6\right)\right]}{\left[\sin\left(\psi + \pi/6\right) - 3\beta_1 \sin\left(\psi - \pi/3\right)\right]}.$$

The value of ξ_c in Eqs. (2.211), (2.212), (2.213), and (2.214) should be eliminated by means of Eq. (2.210).

The procedure for finding the strains is as follows. Let Υ be the value of ρ at which the strains should be calculated at $q = q_m$. It follows from Eq. (2.202) that

$$q_m = 3\beta_1^2 \sin\left(\psi_m + \frac{\pi}{6}\right) + a^2\beta_1 \exp\left[3\beta_1\left(\psi_m - \psi_a\right)\right]\sin\left(\psi_a - \frac{\pi}{3}\right) - 3\beta_0. \tag{2.215}$$

Here ψ_m is the value of ψ_c at $q = q_m$. Equation (2.215) should be solved for ψ_m numerically. Then, the value of ρ_c at $q = q_m$ denoted by ρ_m is determined from Eq. (2.200) as

$$\rho_m = a \exp\left[\frac{3\beta_1}{2}\left(\psi_m - \psi_a\right)\right]\sqrt{\frac{\sin\left(\pi/3 - \psi_a\right)}{\sin\left(\pi/3 - \psi_m\right)}}. \tag{2.216}$$

In the elastic region, $\rho_m \le \Upsilon \le 1$, the distributions of the strains follow from Eq. (1.60) in the form

$$\frac{\varepsilon_r^e}{k} = \frac{\varepsilon_r}{k} = \frac{A_m\left(1+\nu\right)}{\Upsilon^2} + B_m\left(1-\nu\right), \tag{2.217}$$

$$\frac{\varepsilon_\theta^e}{k} = \frac{\varepsilon_\theta}{k} = -\frac{A_m\left(1+\nu\right)}{\Upsilon^2} + B_m\left(1-\nu\right), \quad \frac{\varepsilon_z^e}{k} = \frac{\varepsilon_z}{k} = -2\nu B_m.$$

Having found ψ_m from Eq. (2.215) the values of A_m and B_m are determined by means of Eq. (2.208) as

$$A_m = \frac{\sqrt{3}\beta_1 a^2}{2}\exp\left(3\beta_1\psi_m\right), \quad B_m = 3\left[\beta_0 - \beta_1^2 \sin\left(\psi_m + \frac{\pi}{6}\right)\right]. \tag{2.218}$$

In order to calculate the strains in the plastic region, $a \le \Upsilon \le \rho_m$, it is convenient to introduce the value of ψ_c at $\rho_c = \Upsilon$. This value of ψ_c is denoted by ψ_Υ. The equation for ψ_Υ follows from Eq. (2.200) in the form

$$\Upsilon = a\exp\left[\frac{3\beta_1}{2}\left(\psi_\Upsilon - \psi_a\right)\right]\sqrt{\frac{\sin\left(\pi/3 - \psi_a\right)}{\sin\left(\pi/3 - \psi_\Upsilon\right)}}. \tag{2.219}$$

This equation should be solved for ψ_Υ numerically. Then, the corresponding values of $A = A_\Upsilon$ and $B = B_\Upsilon$ are found from Eq. (2.208) as

$$\begin{aligned} A_\Upsilon &= a^2\beta_1 \exp\left[3\beta_1\left(\psi_\Upsilon - \psi_a\right)\right]\sin\left(\frac{\pi}{3} - \psi_a\right), \\ B_\Upsilon &= 3\beta_0 - 3\beta_1^2 \sin\left(\psi_\Upsilon + \frac{\pi}{6}\right). \end{aligned} \tag{2.220}$$

The elastic portions of the strains are determined from Eq. (1.60) in the form

$$\begin{aligned} \frac{\varepsilon_r^e}{k} &= \frac{A_\Upsilon\left(1+\nu\right)}{\Upsilon^2} + B_\Upsilon\left(1-\nu\right), \\ \frac{\varepsilon_\theta^e}{k} &= -\frac{A_\Upsilon\left(1+\nu\right)}{\Upsilon^2} + B_\Upsilon\left(1-\nu\right), \quad \frac{\varepsilon_z^e}{k} = -2\nu B_\Upsilon. \end{aligned} \tag{2.221}$$

The plastic portions are given by

$$\varepsilon_r^p = \int_{\psi_\Upsilon}^{\psi_m} \xi_r^p d\psi_c, \quad \varepsilon_\theta^p = \int_{\psi_\Upsilon}^{\psi_m} \xi_\theta^p d\psi_c, \quad \varepsilon_z^p = \int_{\psi_\Upsilon}^{\psi_m} \xi_z^p d\psi_c. \tag{2.222}$$

Consider plastically incompressible material first. Substituting Eqs. (2.211) and (2.213) into Eq. (2.222) and taking into account that ψ is independent of ψ_c yield

$$\begin{aligned} \varepsilon_r^p = &\left[\frac{\beta_1\left(\sqrt{3}-\beta_1\right)\cos\psi_\Upsilon - \beta_1\left(1+\sqrt{3}\beta_1\right)\sin\psi_\Upsilon + 2\beta_0}{\beta_1\left(1-\sqrt{3}\beta_1\right)\sin\psi_\Upsilon - \beta_1\left(\sqrt{3}+\beta_1\right)\cos\psi_\Upsilon + 2\beta_0}\right] \times \\ &\times \int_{\psi_\Upsilon}^{\psi_m} \xi_c \exp\left[-\beta_1 \Lambda\left(\psi_\Upsilon, \psi_c\right)\right] d\psi_c, \end{aligned}$$

$$\varepsilon_\theta^p = \int_{\psi_\Upsilon}^{\psi_m} \xi_c \exp\left[-\beta_1 \Lambda\left(\psi_\Upsilon,\ \psi_c\right)\right] d\psi_c, \tag{2.223}$$

$$\varepsilon_z^p = 2\left[\frac{\beta_1^2\left(\cos\psi_\Upsilon + \sqrt{3}\sin\psi_\Upsilon\right) - 2\beta_0}{\beta_1\left(1-\sqrt{3}\beta_1\right)\sin\psi_\Upsilon - \beta_1\left(\sqrt{3}+\beta_1\right)\cos\psi_\Upsilon + 2\beta_0}\right]$$
$$\times \int_{\psi_\Upsilon}^{\psi_m} \xi_c \exp\left[-\beta_1 \Lambda\left(\psi_\Upsilon,\ \psi_c\right)\right] d\psi_c.$$

Here ξ_c should be eliminated by means of Eq. (2.210). Using the solution to Eq. (2.219) and Eq. (2.220) the total strains in the plastic zone are determined from Eqs. (1.3), (2.221) and (2.223) at any given value of Υ.

Consider plastically compressible material. Substituting Eqs. (2.210) and (2.212) into Eq. (2.222) for ε_θ^p and integrating yield

$$\varepsilon_\theta^p = \frac{3\beta_1^2}{2\left(1+9\beta_1^2\right)}\left[1-\nu-\frac{\sqrt{3}\left(1+\nu\right)\exp\left(3\beta_1\psi_a\right)}{2\sin\left(\psi_a-\pi/3\right)}\right]\times \tag{2.224}$$
$$\times\left\{\begin{array}{c}\cos\psi_\Upsilon - 3\beta_1\sin\psi_\Upsilon + \sqrt{3}\left(3\beta_1\cos\psi_\Upsilon + \sin\psi_\Upsilon\right) - \\ -\left[\cos\psi_m - 3\beta_1\sin\psi_m + \sqrt{3}\left(3\beta_1\cos\psi_m + \sin\psi_m\right)\right]\exp\left[3\beta_1\left(\psi_m-\psi_\Upsilon\right)\right]\end{array}\right\}.$$

Substituting Eq. (2.214) into Eq. (2.222), taking into account that ψ is independent of ψ_c and integrating give

$$\varepsilon_r^p = \varepsilon_\theta^p \frac{\left[3\beta_1\sin\left(\psi_\Upsilon-\pi/3\right)+\sin\left(\psi_\Upsilon+\pi/6\right)\right]}{\left[\sin\left(\psi_\Upsilon+\pi/6\right)-3\beta_1\sin\left(\psi_\Upsilon-\pi3\right)\right]}, \tag{2.225}$$
$$\varepsilon_z^p = -\frac{2\varepsilon_\theta^p\beta_1^2}{3}\frac{\left[9\alpha+\left(9+2\alpha^2\right)\sin\left(\psi_\Upsilon+\pi/6\right)\right]}{\left[\sin\left(\psi_\Upsilon+\pi/6\right)-3\beta_1\sin\left(\psi_\Upsilon-\pi/3\right)\right]}.$$

Here the strain ε_θ^p can be eliminated by means of Eq. (2.224). Using the solution to Eqs. (2.219) and (2.220) the total strains in the plastic zone are determined from Eqs. (1.3), (2.221), (2.224), and (2.225) at any given value of Υ.

References

1. Hill R (1950) The mathematical theory of plasticity. Clarendon Press, Oxford
2. Mendelson A (1968) Plasticity: theory and application. The Macmillan Company, New York
3. Chakrabarty J (1987) Theory of plasticity. McGraw-Hill Book Company, New York
4. Rees DWA (2006) Basic engineering plasticity. Elsevier, Amsterdam
5. Masri R, Cohen T, Durban D (2010) Enlargement of a circular hole in a thin plastic sheet: Taylor-Bethe controversy in retrospect. Q J Mech Appl Math 63:589–616

6. Bouvier S, Teodosiu C, Haddadi H, Tabakaru V (2002) Anisotropic work-hardening behavior of structural steels and aluminium alloys at large strains. In: Cescotto S (ed) Proceedings of 6th European mechanics of materials conference. University of Liege, pp 329–326
7. Wu PD, Jain M, Savoie J, MacEwen SR, Tugcu P, Neale KW (2003) Evaluation of anisotropic yield functions for aluminum sheets. Int J Plast 19:121–138
8. Spitzig WA, Sober RJ, Richmond O (1976) The effect of hydrostatic pressure on the deformation behavior of maraging and HY-80 steels and its implications for plasticity theory. Metall Trans 7A:1703–1710
9. Wilson CD (2002) A critical reexamination of classical metal plasticity. Trans ASME J Appl Mech 69:63–68
10. Liu PS (2006) Mechanical behaviors of porous metals under biaxial tensile loads. Mater Sci Eng 422A:176–183

Chapter 3
Thermal Loading

3.1 Disc Loaded by Uniform Temperature Field

The disc shown in Fig. 1.1 is inserted into a rigid container of radius b_0. The inner radius of the disc is stress free. The disc has no stress at the initial temperature. Thermal expansion caused by a rise of temperature and the constraints imposed on the disc affect the zero-stress state. The distribution of temperature is supposed to be uniform. The boundary conditions are

$$u = 0 \tag{3.1}$$

for $\rho = 1$ and

$$\sigma_r = 0 \tag{3.2}$$

for $\rho = a$. At the stage of purely elastic loading these boundary conditions and the solution given by Eqs. (1.29) and (1.61) combine to result in

$$A = \frac{a^2 \tau}{(1+\nu)\, a^2 + 1 - \nu}, \quad B = -\frac{\tau}{(1+\nu)\, a^2 + 1 - \nu}. \tag{3.3}$$

The purely elastic solution is valid up to the value of $\tau = \tau_e$ at which a plastic zone begins to develop. The value of τ_e is determined from Eq. (3.3) and one of the yield criteria. The corresponding value of the function ψ involved in Eqs. (1.32), (1.47) and (1.53) will be denoted by ψ_e. It is understood here that ψ is calculated at the site of plastic yielding initiation and the plastic zone reduces to a circle at this instant. The solutions considered in this chapter are for elastic/plastic discs. Therefore, it is assumed that $\tau \geq \tau_e$ and there is an elastic/plastic boundary, $\rho = \rho_c$, where $\psi = \psi_c$.

S. Alexandrov, *Elastic/Plastic Discs Under Plane Stress Conditions*,
SpringerBriefs in Computational Mechanics, DOI 10.1007/978-3-319-14580-8_3

3.1.1 Yield Criterion (1.5)

Substituting Eq. (3.3) into Eq. (1.31) leads to

$$\tau_e = \frac{1 - \nu + a^2 (1 + \nu)}{2}. \tag{3.4}$$

Since plastic yielding starts to develop from the inner radius of the disc, the material is elastic in the range $\rho_c \leq \rho \leq 1$. Equations (1.29) and (1.61) are valid in this region. However, A and B are not given by Eq. (3.3). Nevertheless, the radial displacement from Eq. (1.61) should satisfy the boundary condition (3.1). Therefore, using Eq. (1.24)

$$\tau = (1 + \nu) A - (1 - \nu) B. \tag{3.5}$$

Equation (1.32) is valid in the plastic zone, $a \leq \rho \leq \rho_c$. Therefore, the radial stress from Eq. (1.32) should satisfy the boundary condition (3.2). Then,

$$\sin \psi_a = 0. \tag{3.6}$$

It is evident that $\sigma_\theta < 0$ at $\rho = a$. It is seen from Eqs. (1.32) and (3.6) that $\sigma_\theta = -\sigma_0 \cos \psi_a$. Therefore, $\cos \psi_a > 0$ and the solution to Eq. (3.6) is

$$\psi_a = 0. \tag{3.7}$$

It is convenient to put $\psi_0 \equiv \psi_a = 0$ and $\rho_0 = a$ in Eq. (1.35). Then, this equation transforms to

$$\rho = \frac{a\sqrt{\sqrt{3}}}{\sqrt{2}\sqrt{\sin(\pi/3 - \psi)}} \exp\left(\frac{\sqrt{3}}{2}\psi\right). \tag{3.8}$$

Using the definition for ψ_c the radius of the elastic/plastic boundary is determined from this equation as

$$\rho_c = \frac{a\sqrt{\sqrt{3}}}{\sqrt{2}\sqrt{\sin(\pi/3 - \psi_c)}} \exp\left(\frac{\sqrt{3}}{2}\psi_c\right). \tag{3.9}$$

Substituting Eqs. (1.29) and (1.32) into Eq. (1.27) yields

$$\frac{A}{\rho_c^2} + B = -\frac{2\sin\psi_c}{\sqrt{3}}, \quad \frac{A}{\rho_c^2} - B = \frac{\sin\psi_c}{\sqrt{3}} + \cos\psi_c. \tag{3.10}$$

Solving these equations for A and B and eliminating ρ_c by means of Eq. (3.9) give

$$A = \frac{a^2}{2} \exp\left(\sqrt{3}\psi_c\right), \quad B = -\frac{\left(\cos\psi_c + \sqrt{3}\sin\psi_c\right)}{2}. \tag{3.11}$$

Substituting Eq. (3.11) into Eq. (3.5) yields

$$\tau = \frac{a^2(1+\nu)}{2} \exp\left(\sqrt{3}\psi_c\right) + \frac{(1-\nu)\left(\cos\psi_c + \sqrt{3}\sin\psi_c\right)}{2}. \tag{3.12}$$

The entire disc becomes plastic when $\rho_c = 1$. The corresponding value of ψ_c is denoted by ψ_q. The equation for determining ψ_q follows from Eq. (3.9) in the form

$$\frac{a\sqrt{\sqrt{3}}}{\sqrt{2}\sqrt{\sin(\pi/3 - \psi_q)}} \exp\left(\frac{\sqrt{3}}{2}\psi_q\right) = 1. \tag{3.13}$$

The corresponding value of τ is denoted by τ_q. The value of τ_q is found from Eq. (3.12) as

$$\tau_q = \frac{a^2(1+\nu)}{2} \exp\left(\sqrt{3}\psi_q\right) + \frac{(1-\nu)\left(\cos\psi_q + \sqrt{3}\sin\psi_q\right)}{2}. \tag{3.14}$$

The distribution of stresses in the elastic zone, $\rho_c \le \rho \le 1$, follows from Eq. (1.29) in which A and B should be eliminated by means of Eq. (3.11). The distribution of stresses in the plastic zone, $a \le \rho \le \rho_c$, follows from Eqs. (1.32) and (3.8) in parametric form with ψ being the parameter. It is seen that the solution depends on ψ_c. In order to find the solution at a prescribed value of τ in the range $\tau_e \le \tau \le \tau_q$, it is necessary to solve Eq. (3.12) for ψ_c.

It is convenient to put $p \equiv \psi_c$ in Eq. (1.16). It is seen from Eq. (3.8) that ψ is independent of ψ_c. Therefore, $\xi_r^e = \xi_\theta^e = \xi_z^e = 0$ and

$$\xi_r = \xi_r^p + \xi_r^T, \quad \xi_\theta = \xi_\theta^p + \xi_\theta^T, \quad \xi_z = \xi_z^p + \xi_z^T \tag{3.15}$$

in the plastic zone. Since ψ_0 is independent of p, Eq. (1.71) becomes

$$\frac{\xi_\theta}{k} = \frac{\xi_c}{k} \exp\left[\sqrt{3}(\psi_c - \psi)\right] + \frac{d\tau}{d\psi_c}\left\{1 - \exp\left[\sqrt{3}(\psi_c - \psi)\right]\right\}. \tag{3.16}$$

Substituting Eq. (1.63) into Eq. (3.16) yields

$$\frac{\xi_\theta}{k} = \exp\left[\sqrt{3}(\psi_c - \psi)\right]\left[(1-\nu)\frac{dB}{d\psi_c} - \frac{(1+\nu)}{\rho_c^2}\frac{dA}{d\psi_c}\right] + \frac{d\tau}{d\psi_c}. \tag{3.17}$$

Differentiating Eq. (3.11) yields

$$\frac{dA}{d\psi_c} = \frac{\sqrt{3}a^2}{2}\exp\left(\sqrt{3}\psi_c\right), \quad \frac{dB}{d\psi_c} = \frac{\left(\sin\psi_c - \sqrt{3}\cos\psi_c\right)}{2}. \tag{3.18}$$

Substituting Eqs. (3.9) and (3.18) into Eq. (3.17) gives

$$\frac{\xi_\theta}{k} = \exp\left[\sqrt{3}\left(\psi_c - \psi\right)\right]\left(\sin\psi_c - \sqrt{3}\cos\psi_c\right) + \frac{d\tau}{d\psi_c}. \tag{3.19}$$

The procedure for finding the strains is as follows. Let Υ be the value of ρ at which the strains should be calculated at $\tau = \tau_m$. Putting $\tau = \tau_m$ in Eq. (3.12) leads to

$$\tau_m = \frac{a^2(1+\nu)}{2}\exp\left(\sqrt{3}\psi_m\right) + \frac{(1-\nu)\left(\cos\psi_m + \sqrt{3}\sin\psi_m\right)}{2} \tag{3.20}$$

Here ψ_m is the value of ψ_c at $\tau = \tau_m$. Equation (3.20) should be solved for ψ_m numerically. Then, the value of ρ_c at $\tau = \tau_m$ denoted by ρ_m is determined from Eq. (3.9) as

$$\rho_m = \frac{a\sqrt{\sqrt{3}}}{\sqrt{2}\sqrt{\sin\left(\pi/3 - \psi_m\right)}}\exp\left(\frac{\sqrt{3}}{2}\psi_m\right). \tag{3.21}$$

In the elastic region, $\rho_m \le \Upsilon \le 1$, the distributions of the strains follow from Eq. (1.61) in the form

$$\frac{\varepsilon_r}{k} = \frac{A_m(1+\nu)}{\Upsilon^2} + B_m(1-\nu) + \tau_m, \quad \frac{\varepsilon_\theta}{k} = -\frac{A_m(1+\nu)}{\Upsilon^2} + B_m(1-\nu) + \tau_m, \tag{3.22}$$

$$\frac{\varepsilon_z}{k} = -2\nu B_m + \tau_m.$$

Having found ψ_m from Eq. (3.20) the values of A_m and B_m are determined by means of Eq. (3.11) as

$$A_m = \frac{a^2}{2}\exp\left(\sqrt{3}\psi_m\right), \quad B_m = -\frac{\left(\cos\psi_m + \sqrt{3}\sin\psi_m\right)}{2}. \tag{3.23}$$

In order to calculate the strains in the plastic region, $a \le \Upsilon \le \rho_m$, it is convenient to introduce the value of ψ_c at $\rho_c = \Upsilon$. This value of ψ_c is denoted by ψ_Υ. The equation for ψ_Υ follows from Eq. (3.9) in the form

$$\Upsilon = \frac{a\sqrt{\sqrt{3}}}{\sqrt{2}\sqrt{\sin(\pi/3 - \psi_\Upsilon)}} \exp\left(\frac{\sqrt{3}}{2}\psi_\Upsilon\right). \tag{3.24}$$

This equation should be solved numerically. Then, the corresponding values of $A = A_\Upsilon$ and $B = B_\Upsilon$ are found from Eq. (3.11) as

$$A_\Upsilon = \frac{a^2}{2}\exp\left(\sqrt{3}\psi_\Upsilon\right), \quad B_\Upsilon = -\frac{\left(\cos\psi_\Upsilon + \sqrt{3}\sin\psi_\Upsilon\right)}{2}. \tag{3.25}$$

The elastic portions of the strains are determined from Eq. (1.60) in the form

$$\frac{\varepsilon_r^e}{k} = \frac{A_\Upsilon(1+\nu)}{\Upsilon^2} + B_\Upsilon(1-\nu), \tag{3.26}$$

$$\frac{\varepsilon_\theta^e}{k} = -\frac{A_\Upsilon(1+\nu)}{\Upsilon^2} + B_\Upsilon(1-\nu), \quad \frac{\varepsilon_z^e}{k} = -2\nu B_\Upsilon.$$

The plastic portions are given by

$$\varepsilon_r^p = \int_{\psi_\Upsilon}^{\psi_m} \xi_r^p d\psi_c, \quad \varepsilon_\theta^p = \int_{\psi_\Upsilon}^{\psi_m} \xi_\theta^p d\psi_c, \quad \varepsilon_z^p = \int_{\psi_\Upsilon}^{\psi_m} \xi_z^p d\psi_c. \tag{3.27}$$

It follows from Eqs. (1.21), (1.28), (1.68), (3.15) and (3.19) that

$$\frac{\xi_r^p}{k} = \exp\left[\sqrt{3}(\psi_c - \psi_\Upsilon)\right]\frac{\left(\sin\psi_c - \sqrt{3}\cos\psi_c\right)\sin(\psi_\Upsilon - \pi/6)}{\cos\psi_\Upsilon},$$

$$\frac{\xi_\theta^p}{k} = \exp\left[\sqrt{3}(\psi_c - \psi_\Upsilon)\right]\left(\sin\psi_c - \sqrt{3}\cos\psi_c\right),$$

$$\frac{\xi_z^p}{k} = -\exp\left[\sqrt{3}(\psi_c - \psi_\Upsilon)\right]\frac{\left(\sin\psi_c - \sqrt{3}\cos\psi_c\right)\sin(\psi_\Upsilon + \pi/6)}{\cos\psi_\Upsilon}.$$

Substituting these equations into (3.27) and integrating yield

$$\frac{\varepsilon_r^p}{k} = \sin\left(\psi_\Upsilon - \frac{\pi}{6}\right)\left\{1 - \frac{\cos\psi_m}{\cos\psi_\Upsilon}\exp\left[\sqrt{3}(\psi_m - \psi_\Upsilon)\right]\right\},$$

$$\frac{\varepsilon_\theta^p}{k} = \cos\psi_\Upsilon - \cos\psi_m \exp\left[\sqrt{3}(\psi_m - \psi_\Upsilon)\right], \tag{3.28}$$

$$\frac{\varepsilon_z^p}{k} = -\sin\left(\psi_\Upsilon + \frac{\pi}{6}\right)\left\{1 - \frac{\cos\psi_m}{\cos\psi_\Upsilon}\exp\left[\sqrt{3}(\psi_m - \psi_\Upsilon)\right]\right\}.$$

It has been taken into account here that ψ is independent of ψ_c. The thermal portions of the strains are found from (1.2) and (1.28) as

$$\frac{\varepsilon_r^T}{k} = \frac{\varepsilon_\theta^T}{k} = \frac{\varepsilon_z^T}{k} = \tau_m. \tag{3.29}$$

Using the solution to Eq. (3.24) and Eq. (3.25) the total strains in the plastic zone are determined from Eqs. (1.3), (3.26), (3.28) and (3.29) at any given value of Υ.

3.1.2 Yield Criterion (1.8)

Substituting Eq. (3.3) into Eq. (1.38) yields

$$b_1 = \frac{\tau^2 [1 + \eta_1 (\eta_1 - \eta)]}{\left[(1+\nu) a^2 + 1 - \nu\right]^2} - \eta_1^2, \quad b_2 = -\frac{2a^2\tau^2 \left(\eta_1^2 - 1\right)}{\left[(1+\nu) a^2 + 1 - \nu\right]^2},$$
$$b_3 = \frac{a^4\tau^2 [1 + \eta_1 (\eta_1 + \eta)]}{\left[(1+\nu) a^2 + 1 - \nu\right]^2}. \tag{3.30}$$

It is convenient to consider the cases $\eta_1 \le 1$ and $\eta_1 > 1$ separately. Firstly, it is assume that $\eta_1 \le 1$. In this case $b_2 \ge 0$ and the plastic zone starts to develop from the inner radius of the disc (see Sect. 1.3.3). Then, Eq. (1.37) at $\rho = a$ and Eq. (3.30) combine to give

$$\tau_e = \frac{\eta_1 \left[1 - \nu + a^2 (1 + \nu)\right]}{2}. \tag{3.31}$$

Since plastic yielding starts to develop from the inner radius of the disc, the material is elastic in the range $\rho_c \le \rho \le 1$. Equations (1.29) and (1.61) are valid in this region. However, A and B are not given by Eq. (3.3). Nevertheless, the radial displacement from Eq. (1.61) should satisfy the boundary condition (3.1). Therefore, using Eq. (1.24)

$$\tau = (1 + \nu) A - (1 - \nu) B. \tag{3.32}$$

Equation (1.47) is valid in the plastic zone, $a \le \rho \le \rho_c$. Therefore, the radial stress from Eq. (1.47) should satisfy the boundary condition (3.2). Then,

$$\sin \psi_a = 0. \tag{3.33}$$

It is evident that $\sigma_\theta < 0$ at $\rho = a$. It is seen from Eqs. (1.9), (1.47) and (3.33) that $\sigma_\theta = -\sigma_0 \eta_1 \cos \psi_a$ at $\rho = a$. Therefore, $\cos \psi_a > 0$ and the solution to Eq. (3.33) is

$$\psi_a = 0. \tag{3.34}$$

It is convenient to put $\psi_0 \equiv \psi_a = 0$ and $\rho_0 = a$ in Eq. (1.50). Therefore, this equation transforms to

$$\ln\left(\frac{\rho}{a}\right) = \frac{\eta_1\sqrt{4-\eta^2}}{2\left[1+\eta_1\left(\eta_1-\eta\right)\right]}\psi + \\ + \frac{(2-\eta\eta_1)}{2\left[1+\eta_1\left(\eta_1-\eta\right)\right]}\ln\left[\frac{\eta_1\sqrt{4-\eta^2}}{\eta_1\sqrt{4-\eta^2}\cos\psi-(2-\eta\eta_1)\sin\psi}\right]. \tag{3.35}$$

Then, using the definition for ψ_c the radius of the elastic/plastic boundary is given by

$$\ln\left(\frac{\rho_c}{a}\right) = \frac{\eta_1\sqrt{4-\eta^2}}{2\left[1+\eta_1\left(\eta_1-\eta\right)\right]}\psi_c + \\ + \frac{(2-\eta\eta_1)}{2\left[1+\eta_1\left(\eta_1-\eta\right)\right]}\ln\left[\frac{\eta_1\sqrt{4-\eta^2}}{\eta_1\sqrt{4-\eta^2}\cos\psi_c-(2-\eta\eta_1)\sin\psi_c}\right]. \tag{3.36}$$

Substituting Eqs. (1.29) and (1.47) into Eq. (1.27) and using Eq. (1.9) yield

$$\frac{A}{\rho_c^2} + B = -\frac{2\sin\psi_c}{\sqrt{4-\eta^2}}, \quad \frac{A}{\rho_c^2} - B = \eta_1\left(\frac{\eta\sin\psi_c}{\sqrt{4-\eta^2}} + \cos\psi_c\right). \tag{3.37}$$

Solving these equations for A and B gives

$$A = \rho_c^2\left[\frac{\eta_1}{2}\left(\frac{\eta\sin\psi_c}{\sqrt{4-\eta^2}} + \cos\psi_c\right) - \frac{\sin\psi_c}{\sqrt{4-\eta^2}}\right], \\ B = -\frac{\eta_1}{2}\left(\frac{\eta\sin\psi_c}{\sqrt{4-\eta^2}} + \cos\psi_c\right) - \frac{\sin\psi_c}{\sqrt{4-\eta^2}}. \tag{3.38}$$

Substituting Eq. (3.38) into Eq. (3.32) yields

$$\tau = (1+\nu)\,\rho_c^2\left[\frac{\eta_1}{2}\left(\frac{\eta\sin\psi_c}{\sqrt{4-\eta^2}} + \cos\psi_c\right) - \frac{\sin\psi_c}{\sqrt{4-\eta^2}}\right] \\ + (1-\nu)\left[\frac{\eta_1}{2}\left(\frac{\eta\sin\psi_c}{\sqrt{4-\eta^2}} + \cos\psi_c\right) + \frac{\sin\psi_c}{\sqrt{4-\eta^2}}\right]. \tag{3.39}$$

Eliminating ρ_c in Eqs. (3.38) and (3.39) by means of Eq. (3.36) supplies A, B and τ as functions of ψ_c. The entire disc becomes plastic when $\rho_c = 1$. The corresponding value of ψ_c is denoted by ψ_q. The equation for determining ψ_q follows from Eq. (3.36) in the form

$$
-\ln a = \frac{\eta_1\sqrt{4-\eta^2}}{2\left[1+\eta_1\left(\eta_1-\eta\right)\right]}\psi_q + \\
+\frac{\left(2-\eta\eta_1\right)}{2\left[1+\eta_1\left(\eta_1-\eta\right)\right]}\ln\left[\frac{\eta_1\sqrt{4-\eta^2}}{\eta_1\sqrt{4-\eta^2}\cos\psi_q-\left(2-\eta\eta_1\right)\sin\psi_q}\right]. \tag{3.40}
$$

The corresponding value of τ is denoted by τ_q. The value of τ_q is found from Eq. (3.39) as

$$
\tau_q = \eta_1\left(\frac{\eta\sin\psi_q}{\sqrt{4-\eta^2}}+\cos\psi_q\right)-\frac{2\nu\sin\psi_q}{\sqrt{4-\eta^2}}. \tag{3.41}
$$

The distribution of stresses in the elastic zone, $\rho_c \le \rho \le 1$, follows from Eq. (1.29) in which A and B should be eliminated by means of Eq. (3.38). The distribution of stresses in the plastic zone, $a \le \rho \le \rho_c$, follows from Eqs. (1.9), (1.47) and (3.35) in parametric form with ψ being the parameter. It is seen that the solution depends on ψ_c. In order to find the solution at a prescribed value of τ in the range $\tau_e \le \tau \le \tau_q$, it is necessary to solve Eq. (3.39) in which ρ_c is eliminated by means of Eq. (3.36) for ψ_c.

It is convenient to put $p \equiv \psi_c$ in Eq. (1.16). It is seen from Eq. (3.35) that ψ is independent of ψ_c. Therefore, $\xi_r^e = \xi_\theta^e = \xi_z^e = 0$ and

$$
\xi_r = \xi_r^P + \xi_r^T, \quad \xi_\theta = \xi_\theta^P + \xi_\theta^T, \quad \xi_z = \xi_z^P + \xi_z^T \tag{3.42}
$$

in the plastic zone. Taking into account that ψ_0 is independent of p and using Eqs. (1.58) and (3.42) it is possible to transform Eq. (1.80) to $\partial\xi_\theta^P/\partial\psi = W_0\left(\psi\right)\xi_\theta^P$. The solution to this equation satisfying the boundary condition (1.64) is

$$
\frac{\xi_\theta^P}{k} = \frac{\xi_c^P}{k}\exp\left[\int_{\psi_c}^{\psi} W_0\left(\mu\right)d\mu\right]. \tag{3.43}
$$

It has been taken into account here that $\xi_\theta^e = 0$ in the plastic zone. The function $W_0\left(\mu\right)$ involved in Eq. (3.43) should be eliminated by means of Eq. (1.81). Equation (1.65) becomes

$$
\frac{\xi_c^P}{k} = -\frac{\left(1+\nu\right)}{\rho_c^2}\frac{dA}{d\psi_c} + \left(1-\nu\right)\frac{dB}{d\psi_c}. \tag{3.44}
$$

Using Eqs. (3.36), (3.38) and (3.44) the value of ξ_c^P involved in Eq. (3.43) is represented as a function of ψ_c.

The procedure for finding the strains is as follows. Let Υ be the value of ρ at which the strains should be calculated at $\tau = \tau_m$. Putting $\tau = \tau_m$ in Eq. (3.39) leads to

$$\tau_m = (1+\nu)\,\rho_m^2\left[\frac{\eta_1}{2}\left(\frac{\eta\sin\psi_m}{\sqrt{4-\eta^2}}+\cos\psi_m\right)-\frac{\sin\psi_m}{\sqrt{4-\eta^2}}\right]+ \\ +(1-\nu)\left[\frac{\eta_1}{2}\left(\frac{\eta\sin\psi_m}{\sqrt{4-\eta^2}}+\cos\psi_m\right)+\frac{\sin\psi_m}{\sqrt{4-\eta^2}}\right]. \quad (3.45)$$

Here ψ_m and ρ_m are the values of ψ_c and ρ_c at $\tau = \tau_m$, respectively. It follows from Eq. (3.36) that

$$\ln\left(\frac{\rho_m}{a}\right) = \frac{\eta_1\sqrt{4-\eta^2}}{2\left[1+\eta_1(\eta_1-\eta)\right]}\psi_m \quad (3.46) \\ +\frac{(2-\eta\eta_1)}{2\left[1+\eta_1(\eta_1-\eta)\right]}\ln\left[\frac{\eta_1\sqrt{4-\eta^2}}{\eta_1\sqrt{4-\eta^2}\cos\psi_m-(2-\eta\eta_1)\sin\psi_m}\right].$$

Equations (3.45) and (3.46) should be solved for ψ_m and ρ_m numerically. In the elastic region, $\rho_m \le \Upsilon \le 1$, the distributions of the strains follow from Eq. (1.61) in the form

$$\frac{\varepsilon_r}{k} = \frac{A_m(1+\nu)}{\Upsilon^2}+B_m(1-\nu)+\tau_m, \quad (3.47) \\ \frac{\varepsilon_\theta}{k} = -\frac{A_m(1+\nu)}{\Upsilon^2}+B_m(1-\nu)+\tau_m,\quad \frac{\varepsilon_z}{k} = -2\nu B_m+\tau_m.$$

Having found ψ_m and ρ_m from Eqs. (3.45) and (3.46) the values of A_m and B_m are determined by means of Eq. (3.38) as

$$A_m = \rho_m^2\left[\frac{\eta_1}{2}\left(\frac{\eta\sin\psi_m}{\sqrt{4-\eta^2}}+\cos\psi_m\right)-\frac{\sin\psi_m}{\sqrt{4-\eta^2}}\right], \\ B_m = -\frac{\eta_1}{2}\left(\frac{\eta\sin\psi_m}{\sqrt{4-\eta^2}}+\cos\psi_m\right)-\frac{\sin\psi_m}{\sqrt{4-\eta^2}}. \quad (3.48)$$

In order to calculate the strains in the plastic region, $a \le \Upsilon \le \rho_m$, it is convenient to introduce the value of ψ_c at $\rho_c = \Upsilon$. This value of ψ_c is denoted by ψ_Υ. The equation for ψ_Υ follows from Eq. (3.36) in the form

$$\ln\left(\frac{\Upsilon}{a}\right) = \frac{\eta_1\sqrt{4-\eta^2}}{2\left[1+\eta_1(\eta_1-\eta)\right]}\psi_\Upsilon \quad (3.49) \\ +\frac{(2-\eta\eta_1)}{2\left[1+\eta_1(\eta_1-\eta)\right]}\ln\left[\frac{\eta_1\sqrt{4-\eta^2}}{\eta_1\sqrt{4-\eta^2}\cos\psi_\Upsilon-(2-\eta\eta_1)\sin\psi_\Upsilon}\right].$$

This equation should be solved for ψ_Υ numerically. Then, the corresponding values of $A = A_\Upsilon$ and $B = B_\Upsilon$ are found from Eq. (3.38) as

$$A_\Upsilon = \Upsilon^2 \left[\frac{\eta_1}{2} \left(\frac{\eta \sin \psi_\Upsilon}{\sqrt{4-\eta^2}} + \cos \psi_\Upsilon \right) - \frac{\sin \psi_\Upsilon}{\sqrt{4-\eta^2}} \right],$$
$$B_\Upsilon = -\frac{\eta_1}{2} \left(\frac{\eta \sin \psi_\Upsilon}{\sqrt{4-\eta^2}} + \cos \psi_\Upsilon \right) - \frac{\sin \psi_\Upsilon}{\sqrt{4-\eta^2}}. \tag{3.50}$$

The elastic portions of the strains are determined from Eq. (1.60) in the form

$$\frac{\varepsilon_r^e}{k} = \frac{A_\Upsilon (1+\nu)}{\Upsilon^2} + B_\Upsilon (1-\nu), \tag{3.51}$$
$$\frac{\varepsilon_\theta^e}{k} = -\frac{A_\Upsilon (1+\nu)}{\Upsilon^2} + B_\Upsilon (1-\nu), \quad \frac{\varepsilon_z^e}{k} = -2\nu B_\Upsilon.$$

The plastic portions are given by

$$\varepsilon_r^p = \int_{\psi_\Upsilon}^{\psi_m} \xi_r^p \, d\psi_c, \quad \varepsilon_\theta^p = \int_{\psi_\Upsilon}^{\psi_m} \xi_\theta^p \, d\psi_c, \quad \varepsilon_z^p = \int_{\psi_\Upsilon}^{\psi_m} \xi_z^p \, d\psi_c. \tag{3.52}$$

Substituting Eq. (3.43) into Eq. (3.52) for ε_θ^p yields

$$\frac{\varepsilon_\theta^p}{k} = \int_{\psi_\Upsilon}^{\psi_m} \frac{\xi_c^p (\mu_1)}{k} \exp \left[\int_{\mu_1}^{\psi_\Upsilon} W_0 (\mu) \, d\mu \right] d\mu_1. \tag{3.53}$$

Here ξ_c^p is represented as a function of μ_1 by means of Eq. (3.44). Substituting Eq. (1.78) into Eq. (3.52), taking into account that ψ is independent of ψ_c and integrating give

$$\varepsilon_r^p = \frac{\eta_1}{2} \left(\sqrt{4-\eta^2} \tan \psi_\Upsilon - \eta \right) \varepsilon_\theta^p, \tag{3.54}$$
$$\varepsilon_z^p = -\frac{\left(2 - \eta\eta_1 + \eta_1 \sqrt{4-\eta^2} \tan \psi_\Upsilon \right)}{2} \varepsilon_\theta^p.$$

Eliminating in these equations ε_θ^p by means of Eq. (3.53) supplies ε_r^p and ε_z^p as functions of ψ_Υ and ψ_m. The thermal portions of the strains are found from (1.2) and (1.28) as

$$\frac{\varepsilon_r^T}{k} = \frac{\varepsilon_\theta^T}{k} = \frac{\varepsilon_z^T}{k} = \tau_m. \tag{3.55}$$

Using the solution to Eq. (3.49) and Eq. (3.50) the total strains in the plastic zone are determined from Eqs. (1.3), (3.51), (3.53), (3.54), and (3.55) at any given value of Υ.

It is now assumed that $\eta_1 > 1$. The case corresponding to Eq. (1.44) is treated in the same manner as the case $\eta_1 \leq 1$ since the plastic zone starts to develop from the inner radius of the disc. However, another plastic zone may start to develop from the outer radius of the disc if τ is large enough. Substituting Eq. (3.38) in which ρ_c is eliminated by means of Eq. (3.36) into Eq. (1.36) at $\rho = 1$ supplies the equation for ψ_c at which the second plastic zone starts to develop. The corresponding value of τ is determined from Eqs. (3.36) and (3.39). The equation for ψ_c may have no solution that predicts that the value of τ is in the range $\tau_e < \tau < \tau_p$. In this case, the second plastic zone does not appear. A solution with two plastic zones is beyond the scope of the present monograph.

Assume that Eq. (1.45) is satisfied. Then, the plastic zone starts to develop from the outer radius of the disc. Equations (1.36) at $\rho = 1$ and (3.3) combine to give

$$\tau_e = \frac{\eta_1 \left[(1+\nu)\, a^2 + 1 - \nu\right]}{\sqrt{1 + \eta_1 (\eta_1 - \eta) - 2a^2 \left(\eta_1^2 - 1\right) + a^4 \left[1 + \eta_1 (\eta_1 + \eta)\right]}}. \tag{3.56}$$

Since plastic yielding starts to develop from the outer radius of the disc, the material is elastic in the range $a \leq \rho \leq \rho_c$. Equations (1.29) and (1.61) are valid in this region. However, A and B are not given by Eq. (3.3). Nevertheless, the radial stress from Eq. (1.29) should satisfy the boundary condition (3.2). Therefore,

$$A = -Ba^2. \tag{3.57}$$

Equation (1.47) are valid in the plastic zone, $\rho_c \leq \rho \leq 1$. Substituting these equations and Eq. (1.29) into Eq. (1.27) and using Eqs. (1.9) and (3.57) yield

$$\frac{2\sin\psi_c}{\sqrt{4-\eta^2}} = A\left(\frac{1}{a^2} - \frac{1}{\rho_c^2}\right), \quad \eta_1\left(\frac{\eta\sin\psi_c}{\sqrt{4-\eta^2}} + \cos\psi_c\right) = A\left(\frac{1}{\rho_c^2} + \frac{1}{a^2}\right). \tag{3.58}$$

Solving these equation for A and ρ_c gives

$$\rho_c^2 = a^2 \frac{\eta_1\left(\eta + \sqrt{4-\eta^2}\cot\psi_c\right) + 2}{\eta_1\left(\eta + \sqrt{4-\eta^2}\cot\psi_c\right) - 2}, \tag{3.59}$$
$$A = \frac{a^2}{2}\left[\frac{(\eta\eta_1 + 2)\sin\psi_c}{\sqrt{4-\eta^2}} + \eta_1\cos\psi_c\right].$$

Let ψ_b be the value of ψ at $\rho = 1$. It is convenient to put $\psi_0 = \psi_b$ and $\rho_0 = 1$ in Eq. (1.50). Then, this equation becomes

$$\ln \rho = \frac{\eta_1\sqrt{4-\eta^2}}{2\left[1+\eta_1\left(\eta_1-\eta\right)\right]}\left(\psi-\psi_b\right)+ + \frac{\left(2-\eta\eta_1\right)}{2\left[1+\eta_1\left(\eta_1-\eta\right)\right]}\ln\left[\frac{\eta_1\sqrt{4-\eta^2}\cos\psi_b-\left(2-\eta\eta_1\right)\sin\psi_b}{\eta_1\sqrt{4-\eta^2}\cos\psi-\left(2-\eta\eta_1\right)\sin\psi}\right]. \tag{3.60}$$

It follows from this equation and the definition for ψ_c that

$$\ln \rho_c^2 = \frac{\eta_1\sqrt{4-\eta^2}}{\left[1+\eta_1\left(\eta_1-\eta\right)\right]}\left(\psi_c-\psi_b\right)+ + \frac{\left(2-\eta\eta_1\right)}{\left[1+\eta_1\left(\eta_1-\eta\right)\right]}\ln\left[\frac{\eta_1\sqrt{4-\eta^2}\cos\psi_b-\left(2-\eta\eta_1\right)\sin\psi_b}{\eta_1\sqrt{4-\eta^2}\cos\psi_c-\left(2-\eta\eta_1\right)\sin\psi_c}\right]. \tag{3.61}$$

Equations (3.59) and (3.61) combine to give

$$2\ln a+\ln\left[\frac{\eta_1\left(\eta+\sqrt{4-\eta^2}\cot\psi_c\right)+2}{\eta_1\left(\eta+\sqrt{4-\eta^2}\cot\psi_c\right)-2}\right] = \frac{\eta_1\sqrt{4-\eta^2}}{\left[1+\eta_1\left(\eta_1-\eta\right)\right]}\left(\psi_c-\psi_b\right)+ + \frac{\left(2-\eta\eta_1\right)}{\left[1+\eta_1\left(\eta_1-\eta\right)\right]}\ln\left[\frac{\eta_1\sqrt{4-\eta^2}\cos\psi_b-\left(2-\eta\eta_1\right)\sin\psi_b}{\eta_1\sqrt{4-\eta^2}\cos\psi_c-\left(2-\eta\eta_1\right)\sin\psi_c}\right]. \tag{3.62}$$

The solution to this equation supplies ψ_c as a function of ψ_b. It is convenient to put $p=\psi_b$ in Eq. (1.16). Differentiating Eqs. (3.57) and (3.59) for A with respect to ψ_b gives

$$\frac{dA}{d\psi_b} = \frac{a^2}{2}\left[\frac{\left(\eta\eta_1+2\right)\cos\psi_c}{\sqrt{4-\eta^2}}-\eta_1\sin\psi_c\right]\frac{d\psi_c}{d\psi_b}, \quad \frac{dB}{d\psi_b} = -\frac{1}{2}\left[\frac{\left(\eta\eta_1+2\right)\cos\psi_c}{\sqrt{4-\eta^2}}-\eta_1\sin\psi_c\right]\frac{d\psi_c}{d\psi_b}. \tag{3.63}$$

The derivative $d\psi_c/d\psi_b$ is found from Eq. (3.62) as

$$\frac{d\psi_c}{d\psi_b} = \frac{V_1\left(\psi_b\right)}{V_2\left(\psi_c\right)} \tag{3.64}$$

where

$$V_1\left(\psi_b\right) = \frac{\left(2-\eta\eta_1\right)}{\left[1+\eta_1\left(\eta_1-\eta\right)\right]}\times \times\left[\frac{\eta_1\sqrt{4-\eta^2}}{2-\eta\eta_1}+\frac{\eta_1\sqrt{4-\eta^2}\sin\psi_b+\left(2-\eta\eta_1\right)\cos\psi_b}{\eta_1\sqrt{4-\eta^2}\cos\psi_b-\left(2-\eta\eta_1\right)\sin\psi_b}\right],$$

$$V_2(\psi_c) = \frac{\eta_1\sqrt{4-\eta^2}}{1+\eta_1(\eta_1-\eta)} + \frac{4\eta_1\sqrt{4-\eta^2}}{4\sin^2\psi_c - \eta_1^2\left(\eta\sin\psi_c + \sqrt{4-\eta^2}\cos\psi_c\right)^2} +$$
$$+ \frac{(2-\eta\eta_1)}{[1+\eta_1(\eta_1-\eta)]}\frac{\left[\eta_1\sqrt{4-\eta^2}\sin\psi_c + (2-\eta\eta_1)\cos\psi_c\right]}{\left[\eta_1\sqrt{4-\eta^2}\cos\psi_c - (2-\eta\eta_1)\sin\psi_c\right]}.$$

Substituting Eqs. (3.57), (3.59) and (3.64) into Eq. (1.63) leads to

$$\frac{\xi_c}{k} = \frac{\left(\eta\eta_1 - 2\nu + \eta_1\sqrt{4-\eta^2}\cot\psi_c\right)}{\sqrt{4-\eta^2}\left(2+\eta\eta_1+\eta_1\sqrt{4-\eta^2}\cot\psi_c\right)} \times$$
$$\times \left[\eta_1\sqrt{4-\eta^2}\sin\psi_c - (2+\eta\eta_1)\cos\psi_c\right]\frac{V_1(\psi_b)}{V_2(\psi_c)} + \frac{d\tau}{d\psi_b}. \tag{3.65}$$

Equation (1.82) supplies the distribution of ξ_θ in the plastic zone as

$$\frac{\xi_\theta}{k} = \left\{\begin{array}{l} \Omega_A(\psi_b)\int\limits_{\psi_c}^{\psi}\exp\left[-\int\limits_{\psi_c}^{\mu_1}W_0(\mu)\,d\mu\right]W_1(\mu_1)\,d\mu_1 - \\ -\frac{d\tau}{d\psi_b}\int\limits_{\psi_c}^{\psi}\exp\left[-\int\limits_{\psi_c}^{\mu_1}W_0(\mu)\,d\mu\right]W_0(\mu_1)\,d\mu_1 + \frac{\xi_c}{k}\end{array}\right\} \times$$
$$\times \exp\left[\int\limits_{\psi_c}^{\psi}W_0(\mu)\,d\mu\right]. \tag{3.66}$$

The boundary condition (3.1) shows that $\xi_\theta = 0$ at $\rho = 1$ (or $\psi = \psi_b$). Then, it follows from Eq. (3.66) that

$$\Omega_A(\psi_b)\int\limits_{\psi_c}^{\psi_b}\exp\left[-\int\limits_{\psi_c}^{\mu_1}W_0(\mu)\,d\mu\right]W_1(\mu_1)\,d\mu_1 -$$
$$-\frac{d\tau}{d\psi_b}\int\limits_{\psi_c}^{\psi_b}\exp\left[-\int\limits_{\psi_c}^{\mu_1}W_0(\mu)\,d\mu\right]W_0(\mu_1)\,d\mu_1 + \frac{\xi_c}{k} = 0. \tag{3.67}$$

In Eqs. (3.66) and (3.67), the functions $\Omega_A(\psi_b)$, $W_0(\psi)$ and $W_1(\psi)$ should be eliminated by means of Eqs. (1.79) and (1.81). Using Eqs. (1.79), (1.81), (3.65) and the solution to Eq. (3.62) it is possible to express the derivative $d\tau/d\psi_b$ as a function of ψ_b using Eq. (3.67). Thus τ can be found as a function of ψ_b by integration with the use of the condition that $\tau = \tau_e$ at $\psi_b = \psi_c = \psi_e$. The equation for ψ_e follows from Eq. (3.59) at $\rho_c = 1$ and $\psi_c = \psi_e$. Then,

$$\cot \psi_e = \frac{\eta\eta_1 \left(a^2 - 1\right) + 2\left(a^2 + 1\right)}{\eta_1 \sqrt{4 - \eta^2}\left(1 - a^2\right)}. \tag{3.68}$$

The unique solution to this equation is determined by comparing A found by means of Eq. (3.59) in which ψ_c should be replaced with ψ_e and A found by means of Eq. (3.3) in which τ should be replaced with τ_e given in Eq. (3.56).

Using Eq. (1.79) it is possible to transform Eqs. (1.83) and (1.84) to

$$\frac{d\psi}{d\psi_b} = \frac{\eta_1\sqrt{4 - \eta^2} - (2 - \eta\eta_1)\tan\psi}{\eta_1\sqrt{4 - \eta^2} - (2 - \eta\eta_1)\tan\psi_b}, \tag{3.69}$$

$$\frac{d\varepsilon_r}{d\psi_b} = \xi_r, \quad \frac{d\varepsilon_\theta}{d\psi_b} = \xi_\theta, \quad \frac{d\varepsilon_z}{d\psi_b} = \xi_z. \tag{3.70}$$

Integrating Eq. (3.69) gives

$$\eta_1\sqrt{4 - \eta^2}\left(\psi - \psi_b\right) + (2 - \eta\eta_1)\ln\left[\frac{\eta_1\sqrt{4 - \eta^2}\cos\psi_b - (2 - \eta\eta_1)\sin\psi_b}{\eta_1\sqrt{4 - \eta^2}\cos\psi - (2 - \eta\eta_1)\sin\psi}\right] =$$
$$= 2\left[1 + \eta_1\left(\eta_1 - \eta\right)\right]\ln C \tag{3.71}$$

where C is constant on each characteristic curve. Comparing Eqs. (3.60) and (3.71) shows that $C = \rho$. Therefore, integrating along the characteristics is equivalent to integrating at fixed values of ρ. Since the derivative $d\tau/d\psi_b$ is found as a function of ψ_b and ψ_c from Eq. (3.67), ξ_c is also found as a function of ψ_b and ψ_c from Eq. (3.65). Then, Eq. (3.66) supplies ξ_θ as a function of ψ, ψ_b and ψ_c. Eliminating ψ_c by means of the solution to Eq. (3.62) gives ξ_θ as a function of ψ and ψ_b. Further eliminating ψ by means of the solution to Eq. (3.71) at any given value of $C = \rho$ determines the right hand side of Eq. (3.70) for ε_θ as a function of ψ_b. This function is denoted by $E_\theta(\psi_b)$. Therefore, Eq. (3.70) for ε_θ can be integrated numerically. In particular, the value of ε_θ at $\psi_b = \psi_m$ and $\rho = C$ is given by

$$\varepsilon_\theta = \int\limits_{\psi_i}^{\psi_m} E_\theta(\psi_b)\, d\psi_b + E_\theta^e. \tag{3.72}$$

Here ψ_m is prescribed and Eq. (3.72) supplies ε_θ in the plastic zone. Since the value of τ has been found as a function of ψ_b, it is possible to calculate the value of τ corresponding to $\psi_b = \psi_m$. This value is denoted by τ_m. The procedure to determine ψ_i and E_θ^e involved in Eq. (3.72) is as follows. The value of E_θ^e is the circumferential strain at $\rho = \rho_c = C$ and the value of ψ_i is the value of ψ_b at $\rho_c = C$. It is seen from

Eq. (1.26)[1] that E_θ^e is determined from the solution in the elastic zone. Alternatively, since the stresses are continuous across the elastic/plastic boundary, the value of E_θ^e can be found from Eqs. (1.1) and (1.47). Let ψ_C be the value of ψ_c at $\rho_c = C$. Then, the equations for ψ_C follows from Eq. (3.59) as

$$C^2 = a^2 \frac{\left[\eta_1 \left(\eta + \sqrt{4-\eta^2}\cot\psi_C\right) + 2\right]}{\left[\eta_1 \left(\eta + \sqrt{4-\eta^2}\cot\psi_C\right) - 2\right]}. \tag{3.73}$$

Having found the value of ψ_C the value of E_θ^e is determined from Eqs. (1.1), (1.9) and (1.47) as

$$E_\theta^e = -\eta_1 \cos\psi_C - \frac{(\eta\eta_1 - 2\nu)}{\sqrt{4-\eta^2}} \sin\psi_C. \tag{3.74}$$

Substituting $\psi_c = \psi_C$ into Eq. (3.62) and solving this equating for ψ_b supply the value of ψ_i involved in Eq. (3.72).

The distributions of ε_r and ε_z in the plastic zone are determined in a similar manner. In particular, using Eqs. (1.22), (1.58) and (1.78)

$$\begin{aligned}
\xi_r &= \xi_r^e + k\frac{d\tau}{d\psi_b} + \xi_r^p = \xi_r^e + k\frac{d\tau}{d\psi_b} + \frac{\xi_\theta^p \eta_1}{2}\left(\sqrt{4-\eta^2}\tan\psi - \eta\right) = \\
&= \xi_r^e + k\frac{d\tau}{d\psi_b} + \frac{\eta_1}{2}\left(\sqrt{4-\eta^2}\tan\psi - \eta\right)\left(\xi_\theta - \xi_\theta^e - k\frac{d\tau}{d\psi_b}\right), \\
\xi_z &= \xi_z^e + k\frac{d\tau}{d\psi_b} + \xi_z^p = \xi_z^e + k\frac{d\tau}{d\psi_b} - \frac{\xi_\theta^p}{2}\left(\sqrt{4-\eta^2}\tan\psi + 2 - \eta\eta_1\right) = \\
&= \xi_z^e + k\frac{d\tau}{d\psi_b} - \frac{1}{2}\left(\sqrt{4-\eta^2}\tan\psi + 2 - \eta\eta_1\right)\left(\xi_\theta - \xi_\theta^e - k\frac{d\tau}{d\psi_b}\right).
\end{aligned} \tag{3.75}$$

Using Eqs. (1.76), (1.79) and (3.66) the right hand sides of Eq. (3.75) are expressed in terms of ψ and ψ_b. Eliminating ψ by means of the solution to Eq. (3.71) at a given value of C determines the right hand sides of these equations as functions of ψ_b. These functions are denoted by $E_r(\psi_b)$ and $E_z(\psi_b)$. Equation (3.70) for ε_r and ε_z can be integrated numerically. In particular,

$$\varepsilon_r = \int_{\psi_i}^{\psi_m} E_r(\psi_b)\, d\psi_b + E_r^e, \quad \varepsilon_z = \int_{\psi_i}^{\psi_m} E_z(\psi_b)\, d\psi_b + E_z^e. \tag{3.76}$$

These equations supply ε_r and ε_z in the plastic zone. In Eq. (3.76), E_r^e and E_z^e are the radial and axial strains, respectively, at $\rho = \rho_c = C$. These strains are determined from Eq. (1.61) at $\rho = \rho_c = C$, (3.57) and (3.59) or from Eqs. (1.1) and (1.47) at $\psi = \psi_c = \psi_C$. As a result,

$$\frac{E_r^e}{k} = \nu\eta_1 \cos\psi_C + \frac{(\nu\eta\eta_1 - 2)}{\sqrt{4-\eta^2}} \sin\psi_C, \tag{3.77}$$
$$\frac{E_z^e}{k} = \nu\left[\eta_1 \cos\psi_C + \frac{(2+\eta\eta_1)}{\sqrt{4-\eta^2}} \sin\psi_C\right].$$

Having found the distributions of the total strains in the plastic zone it is possible to determine their plastic portion by means of Eq. (1.3) in which the elastic and thermal strains should be eliminated using Eqs. (1.1), (1.2), (1.9), (1.28), (1.47), and (3.60). The total strains in the elastic zone follows from Eq. (1.61) in which A and B should be eliminated by means of Eqs. (3.57) and (3.59). The value of ψ_c involved in Eq. (3.59) is determined from Eq. (3.62) assuming that $\psi_b = \psi_m$.

The solution found is restricted by the condition that another plastic zone can appear at the inner radius of the disc. This condition follows from Eq. (1.36) at $\rho = a$. Eliminating A and B by means of Eqs. (3.57) and (3.59) yields

$$\left|\frac{(\eta\eta_1 + 2)\sin\psi_c}{\sqrt{4-\eta^2}} + \eta_1\cos\psi_c\right| \le \eta_1. \tag{3.78}$$

If this inequality is not satisfied then it is necessary to find the solution with two plastic zones. This solution is beyond the scope of the present monograph. So is the solution for discs satisfying Eq. (1.46). In this case two plastic zones start to develop from the inner and outer radii of the disc simultaneously.

3.1.3 Yield Criterion (1.11)

Substituting Eq. (3.3) into Eq. (1.52) leads to

$$\tau_e = \frac{3\left[1 - \nu + a^2(1+\nu)\right]}{2(3-\alpha)}. \tag{3.79}$$

Since plastic yielding starts to develop from the inner radius of the disc, the material is elastic in the range $\rho_c \le \rho \le 1$. Equations (1.29) and (1.61) are valid in this region. However, A and B are not given by Eq. (3.3). Nevertheless, the radial displacement from Eq. (1.61) should satisfy the boundary condition (3.1). Therefore, using Eq. (1.24)

$$\tau = (1+\nu)A - (1-\nu)B. \tag{3.80}$$

Equation (1.53) is valid in the plastic zone, $a \le \rho \le \rho_c$. Therefore, the radial stress from Eq. (1.53) should satisfy the boundary condition (3.2). Then,

$$3\beta_0 - \frac{\beta_1}{2}\left(1 + 3\sqrt{3}\beta_1\right)\sin\psi_a + \frac{\sqrt{3}\beta_1}{2}\left(1 - \sqrt{3}\beta_1\right)\cos\psi_a = 0. \quad (3.81)$$

It is evident that $\sigma_\theta < 0$ at $\rho = a$. Therefore, it follows from Eq. (1.53) that the unique solution to Eq. (3.81) is found by using the condition

$$3\beta_0 + \frac{\beta_1}{2}\left(1 - 3\sqrt{3}\beta_1\right)\sin\psi_a - \frac{\sqrt{3}\beta_1}{2}\left(1 + \sqrt{3}\beta_1\right)\cos\psi_a < 0. \quad (3.82)$$

In particular, Eqs. (3.81) and (3.82) combine to give $\sin(\pi/3 - \psi_a) > 0$ or

$$-\frac{2\pi}{3} < \psi_a < \frac{4\pi}{3}. \quad (3.83)$$

It is convenient to put $\psi_0 \equiv \psi_a$ and $\rho_0 = a$ in Eq. (1.57). Therefore, this equation transforms to

$$\rho = a\exp\left[\frac{3\beta_1}{2}(\psi - \psi_a)\right]\sqrt{\frac{\sin(\psi_a - \pi/3)}{\sin(\psi - \pi/3)}}. \quad (3.84)$$

Then, using the definition for ψ_c the radius of the elastic/plastic boundary is given by

$$\rho_c = a\exp\left[\frac{3\beta_1}{2}(\psi_c - \psi_a)\right]\sqrt{\frac{\sin(\psi_a - \pi/3)}{\sin(\psi_c - \pi/3)}}. \quad (3.85)$$

Substituting Eqs. (1.29) and (1.53) into Eq. (1.27) yields

$$\frac{A}{\rho_c^2} + B = 3\beta_0 - \frac{\beta_1}{2}\left(1 + 3\sqrt{3}\beta_1\right)\sin\psi_c + \frac{\sqrt{3}\beta_1}{2}\left(1 - \sqrt{3}\beta_1\right)\cos\psi_c, \quad (3.86)$$

$$-\frac{A}{\rho_c^2} + B = 3\beta_0 + \frac{\beta_1}{2}\left(1 - 3\sqrt{3}\beta_1\right)\sin\psi_c - \frac{\sqrt{3}\beta_1}{2}\left(1 + \sqrt{3}\beta_1\right)\cos\psi_c.$$

Solving these equations for A and B and eliminating ρ_c by means of Eq. (3.85) give

$$A = a^2\beta_1\exp\left[3\beta_1(\psi_c - \psi_a)\right]\sin\left(\frac{\pi}{3} - \psi_a\right), \quad (3.87)$$

$$B = 3\left[\beta_0 - \beta_1^2\cos\left(\frac{\pi}{3} - \psi_c\right)\right].$$

Substituting Eq. (3.87) into Eq. (3.80) yields

$$\tau = a^2 (1+\nu) \beta_1 \exp[3\beta_1 (\psi_c - \psi_a)] \sin\left(\frac{\pi}{3} - \psi_a\right) - \\ - 3(1-\nu)\left[\beta_0 - \beta_1^2 \cos\left(\frac{\pi}{3} - \psi_c\right)\right]. \quad (3.88)$$

The entire disc becomes plastic when $\rho_c = 1$. The corresponding value of ψ_c is denoted by ψ_q. The equation for determining ψ_q follows from Eq. (3.85) in the form

$$a \exp\left[\frac{3\beta_1}{2}(\psi_q - \psi_a)\right]\sqrt{\frac{\sin(\psi_a - \pi/3)}{\sin(\psi_q - \pi/3)}} = 1. \quad (3.89)$$

The corresponding value of τ is denoted by τ_q. The value of τ_q is found from Eq. (3.88) as

$$\tau_q = a^2 (1+\nu) \beta_1 \exp[3\beta_1 (\psi_q - \psi_a)] \sin\left(\frac{\pi}{3} - \psi_a\right) - \\ - 3(1-\nu)\left[\beta_0 - \beta_1^2 \cos\left(\frac{\pi}{3} - \psi_q\right)\right]. \quad (3.90)$$

The distribution of stresses in the elastic zone, $\rho_c \le \rho \le 1$, follows from Eq. (1.29) in which A and B should be eliminated by means of Eq. (3.87). The distribution of stresses in the plastic zone, $a \le \rho \le \rho_c$, follows from Eqs. (1.53) and (3.84) in parametric form with ψ being the parameter. It is seen that the solution depends on ψ_c. In order to find the solution at a prescribed value of τ in the range $\tau_e \le \tau \le \tau_q$, it is necessary to solve Eq. (3.88) for ψ_c.

It is convenient to put $p \equiv \psi_c$ in Eq. (1.16). It is seen from Eqs. (3.81) and (3.84) that ψ is independent of ψ_c. Therefore, $\xi_r^e = \xi_\theta^e = \xi_z^e = 0$ and

$$\xi_r = \xi_r^p + \xi_r^T, \quad \xi_\theta = \xi_\theta^p + \xi_\theta^T, \quad \xi_z = \xi_z^p + \xi_z^T \quad (3.91)$$

in the plastic zone. Taking into account that ψ_0 is independent of p and using Eqs. (1.58) and (3.91) it is possible to transform Eq. (1.88) to $\partial\xi_\theta^p/\partial\psi = W_0(\psi)\,\xi_\theta^p$. The solution to this equation satisfying the boundary condition (1.64) is

$$\frac{\xi_\theta^p}{k} = \frac{\xi_c^p}{k}\exp\left[\int_{\psi_c}^{\psi} W_0(\mu)\,d\mu\right]. \quad (3.92)$$

It has been taken into account here that $\xi_\theta^e = 0$ in the plastic zone. The function $W_0(\mu)$ involved in Eq. (3.92) should be eliminated by means of Eq. (1.89). Equation (3.92) is valid for plastically incompressible materials. The corresponding equation for plastically compressible materials follows from Eq. (1.94) in the form

$$\frac{\xi_\theta}{k} = \frac{\xi_c^p}{k} \exp\left[3\beta_1\left(\psi_c - \psi\right)\right]. \tag{3.93}$$

Equation (1.65) becomes

$$\frac{\xi_c^p}{k} = -\frac{(1+\nu)}{\rho_c^2}\frac{dA}{d\psi_c} + (1-\nu)\frac{dB}{d\psi_c}.$$

Substituting Eqs. (3.85) and (3.87) into this equation gives

$$\frac{\xi_c^p}{k} = 6\beta_1^2 \sin\left(\psi_c - \frac{\pi}{3}\right). \tag{3.94}$$

The procedure for finding the strains is as follows. Let Υ be the value of ρ at which the strains should be calculated at $\tau = \tau_m$. Putting $\tau = \tau_m$ in Eq. (3.88) leads to

$$\tau_m = a^2(1+\nu)\beta_1 \exp\left[3\beta_1\left(\psi_m - \psi_a\right)\right]\sin\left(\frac{\pi}{3} - \psi_a\right) - \\ - 3(1-\nu)\left[\beta_0 - \beta_1^2\cos\left(\frac{\pi}{3} - \psi_m\right)\right]. \tag{3.95}$$

Here ψ_m is the value of ψ_c at $\tau = \tau_m$. Equation (3.95) should be solved for ψ_m numerically. Then, the value of ρ_c at $\tau = \tau_m$ denoted by ρ_m is determined from Eq. (3.85) as

$$\rho_m = a\exp\left[\frac{3\beta_1}{2}\left(\psi_m - \psi_a\right)\right]\sqrt{\frac{\sin\left(\psi_a - \pi/3\right)}{\sin\left(\psi_m - \pi/3\right)}}. \tag{3.96}$$

In the elastic region, $\rho_m \le \Upsilon \le 1$, the distributions of the strains follow from Eq. (1.61) in the form

$$\frac{\varepsilon_r}{k} = \frac{A_m(1+\nu)}{\Upsilon^2} + B_m(1-\nu) + \tau_m, \tag{3.97}$$

$$\frac{\varepsilon_\theta}{k} = -\frac{A_m(1+\nu)}{\Upsilon^2} + B_m(1-\nu) + \tau_m, \quad \frac{\varepsilon_z}{k} = -2\nu B_m + \tau_m.$$

Having found ψ_m from Eq. (3.95) the values of A_m and B_m are determined by means of Eq. (3.87) as

$$A_m = a^2\beta_1 \exp\left[3\beta_1\left(\psi_m - \psi_a\right)\right]\sin\left(\frac{\pi}{3} - \psi_a\right),$$

$$B_m = 3\left[\beta_0 - \beta_1^2\cos\left(\frac{\pi}{3} - \psi_m\right)\right]. \tag{3.98}$$

In order to calculate the strains in the plastic region, $a \leq \Upsilon \leq \rho_m$, it is convenient to introduce the value of ψ_c at $\rho_c = \Upsilon$. This value of ψ_c is denoted by ψ_Υ. The equation for ψ_Υ follows from Eq. (3.85) in the form

$$\Upsilon = a \exp\left[\frac{3\beta_1}{2}(\psi_\Upsilon - \psi_a)\right]\sqrt{\frac{\sin(\psi_a - \pi/3)}{\sin(\psi_\Upsilon - \pi/3)}}. \tag{3.99}$$

This equation should be solved numerically. Then, the corresponding values of $A = A_\Upsilon$ and $B = B_\Upsilon$ are found from Eq. (3.87) as

$$A_\Upsilon = a^2\beta_1 \exp[3\beta_1(\psi_\Upsilon - \psi_a)]\sin\left(\frac{\pi}{3} - \psi_a\right),$$
$$B_\Upsilon = 3\left[\beta_0 - \beta_1^2\cos\left(\frac{\pi}{3} - \psi_\Upsilon\right)\right]. \tag{3.100}$$

The elastic portions of the strains are determined from Eq. (1.60) in the form

$$\frac{\varepsilon_r^e}{k} = \frac{A_\Upsilon(1+\nu)}{\Upsilon^2} + B_\Upsilon(1-\nu), \tag{3.101}$$
$$\frac{\varepsilon_\theta^e}{k} = -\frac{A_\Upsilon(1+\nu)}{\Upsilon^2} + B_\Upsilon(1-\nu), \quad \frac{\varepsilon_z^e}{k} = -2\nu B_\Upsilon.$$

The plastic portions are given by

$$\varepsilon_r^p = \int_{\psi_\Upsilon}^{\psi_m} \xi_r^p \, d\psi_c, \quad \varepsilon_\theta^p = \int_{\psi_\Upsilon}^{\psi_m} \xi_\theta^p \, d\psi_c, \quad \varepsilon_z^p = \int_{\psi_\Upsilon}^{\psi_m} \xi_z^p \, d\psi_c. \tag{3.102}$$

Consider plastically incompressible material first. Substituting Eq. (3.92) into Eq. (3.102) for ε_θ^p yields

$$\frac{\varepsilon_\theta^p}{k} = \int_{\psi_\Upsilon}^{\psi_m} \frac{\xi_c^p(\mu_1)}{k} \exp\left[\int_{\mu_1}^{\psi_\Upsilon} W_0(\mu)\, d\mu\right] d\mu_1. \tag{3.103}$$

Here ξ_c^p should be eliminated by means of Eq. (3.94). Substituting Eq. (1.86) into Eq. (3.102), taking into account that ψ is independent of ψ_c and integrating give

$$\varepsilon_r^p = \varepsilon_\theta^p \left[\frac{\beta_1\left(\sqrt{3} - \beta_1\right)\cos\psi_\Upsilon - \beta_1\left(1 + \sqrt{3}\beta_1\right)\sin\psi_\Upsilon + 2\beta_0}{\beta_1\left(1 - \sqrt{3}\beta_1\right)\sin\psi_\Upsilon - \beta_1\left(\sqrt{3} + \beta_1\right)\cos\psi_\Upsilon + 2\beta_0}\right], \tag{3.104}$$
$$\varepsilon_z^p = 2\varepsilon_\theta^p \left[\frac{\beta_1^2\left(\cos\psi_\Upsilon + \sqrt{3}\sin\psi_\Upsilon\right) - 2\beta_0}{\beta_1\left(1 - \sqrt{3}\beta_1\right)\sin\psi_\Upsilon - \beta_1\left(\sqrt{3} + \beta_1\right)\cos\psi_\Upsilon + 2\beta_0}\right].$$

Eliminating in these equations ε_θ^p by means of Eq. (3.103) supplies ε_r^p and ε_z^p as functions of ψ_Υ and ψ_m. The thermal portions of the strains are found from (1.2) and (1.28) as

$$\frac{\varepsilon_r^T}{k} = \frac{\varepsilon_\theta^T}{k} = \frac{\varepsilon_z^T}{k} = \tau_m. \tag{3.105}$$

Using the solution to Eqs. (3.99) and (3.100) the total strains in the plastic zone are determined from Eqs. (1.3), (3.101), (3.103), (3.104), and (3.105) at any given value of Υ.

Consider plastically compressible material. Substituting Eq. (3.93) into Eq. (3.102) for ε_θ^p yields

$$\frac{\varepsilon_\theta^p}{k} = \int\limits_{\psi_\Upsilon}^{\psi_m} \frac{\xi_c^p(\psi_c)}{k} \exp\left[3\beta_1(\psi_c - \psi_\Upsilon)\right] d\psi_c. \tag{3.106}$$

Here ξ_c^p should be eliminated by means of Eq. (3.94). Substituting Eq. (1.93) into Eq. (3.102), taking into account that ψ is independent of ψ_c and integrating give

$$\varepsilon_r^p = \varepsilon_\theta^p \left[\frac{3\beta_1 \sin(\psi_\Upsilon - \pi/3) + \sin(\psi_\Upsilon + \pi/6)}{\sin(\psi_\Upsilon + \pi/6) - 3\beta_1 \sin(\psi_\Upsilon - \pi/3)}\right], \tag{3.107}$$
$$\varepsilon_z^p = -\frac{2\beta_1^2 \varepsilon_\theta^p}{3}\left[\frac{9\alpha + \left(9 + 2\alpha^2\right)\sin(\psi_\Upsilon + \pi/6)}{\sin(\psi_\Upsilon + \pi/6) - 3\beta_1 \sin(\psi_\Upsilon - \pi/3)}\right].$$

Eliminating in these equations ε_θ^p by means of Eq. (3.106) supplies ε_r^p and ε_z^p as functions of ψ_Υ and ψ_m. Using the solution to Eqs. (3.99) and (3.100) the total strains in the plastic zone are determined from Eqs. (1.3), (3.101), (3.105), (3.106), and (3.107) at any given value of Υ.

Printed in the United States
By Bookmasters